S0-BDG-935

Energy Prices 1960-73

A Reference Book

A Report to the Energy Policy Project of the Ford Foundation

Energy Prices 1960-73

Prepared by
Foster Associates
Washington, D.C.

Ballinger Publishing Company • Cambridge, Mass.
A Subsidiary of J.B. Lippincott Company

Published in the United States of America by Ballinger Publishing Company, Cambridge, Mass.

Third Printing 1975

Library of Congress Catalog Card Number: 74-9503

International Standard Book Number: 0-88410-327-7 HB
0-88410-330-7 PB

Printed in the United States of America

Library of Congress Cataloging in publication Data

Foster Associates.
Energy prices 1960-73.

I. Petroleum products—Prices—United States.
II. Gas, Natural—United States—Rates.
III. Electric Utilities—United States—Rates. I. Title.
HD9564.F67
ISBN 0-88410-327-7 338.4'3665'5 74-9503

Table of Contents

List of Tables

Part I

Part II

Preface

The Energy Policy Project was initiated by the Ford Foundation in 1971 to explore alternative national energy policies. This book, one of the series of studies commissioned by the Project, is a reference work that compiles information about energy prices. The special reports are being released as they are completed rather than delaying them until the final report of the Energy Policy Project is ready because I believe they can make a timely contribution to the public discussion of energy policies. However, each special report deals with only a part of the energy puzzle; our final report, to be published later in 1974, will attempt to integrate these parts into a comprehensible whole setting forth the nation's energy policy options as we see them.

This book, like the others in the series, has been reviewed by experts in the field not otherwise associated with the Project in order to give due consideration to differing points of view. We have published the outside reviewer comments on pages 257 and 258 of this book.

Energy prices 1960–73 is the authors' report to the Ford Foundation's Energy Policy Project and neither the Foundation, its Energy Policy Project, nor the Project's Advisory necessarily endorse its contents or conclusions. We will express our views in the Project's final report, which will culminate this series of publications.

S. David Freeman
Director
Energy Policy Project

Acknowledgments

This study was prepared by the Energy Division of Foster Associates, Incorporated, under the direction of Radford L. Schantz. Jack D. Colclough, project manager, and Anne Probst, project statistician, contributed to all parts of the study. Henry Herz assisted in structuring and advising in the compilation of the prices of electricity for commercial and industrial use. Wayne Mikutowicz was largely responsible for the compilation and analysis of the coal price data, and William Foster provided assistance in the development of the data on taxes and on the pricing of electricity and gas for central air conditioning. Other staff members assisting in the preparation of the report included Isabel Bowen, Johanna Bower, John Brickhill, Cathi Gordon, Irene Jacobs, Gloria Jones, James Spearman, and Kathy Thomas.

Dr. Kenneth Saulter was most helpful in suggesting guidelines to provide a more useful product to the Energy Policy Project.

Part One
Energy Price Developments In 1973

Chapter One

Introduction

This Part reviews some of the more important developments in energy prices during the year 1973. Part II, the basic report on *Retail and Wholesale Prices for Primary and Secondary Energy Sources in the United States,* covers the period 1960–1972. While energy prices changed moderately during this 13–year period, the changes were relatively small in comparison with the dramatic increases that occurred in 1973.

The definitions and concepts used in this Part of the study are also used in Part II. The energy sources priced are the same and include natural gas, distillate fuel oil, residual fuel oil, motor gasoline, coal, and electricity. While twenty-four Standard Metropolitan Statistical Areas will be covered in Part II, in Part I the number is limited to nine, one for each Census region, as follows: Boston, Massachusetts; New York City; Charlotte, North Carolina; Chicago, Illinois; Nashville, Tennessee; Minneapolis-St. Paul, Minnesota; Houston, Texas; Denver, Colorado; and Los Angeles, California. Moreover, the market sectors for which energy prices were compiled were reduced to four: substitutable and non-substitutable residential baseload, home heat, industrial use, and service stations.

On the other hand, this Part, unlike Part II, includes the pricing of selected crude oils both at the wellhead and delivered to designated refinery centers. The field price of natural gas and the price of coal at the mine are also considered. In addition, in the appendix to this Part are price control chronologies on the petroleum industry, public utilities, and the coal industry, as well as a chronology on actions and agreements of the Organization of Petroleum Exporting Countries (OPEC) having a significant effect on crude oil prices.

OVERVIEW

The rising trends in the prices of practically all forms of energy in recent years accelerated markedly in 1973. A particularly sharp increase in petroleum fuels led the advance in energy prices, but the other fossil fuels—natural gas and coal—also registered appreciable price increases. The gains in the prices of oil, gas, and coal contributed importantly to the large increases in electric rates.

This acceleration in the rise in energy prices was not a unique price development in the nation's economy in 1973, but part of the accelerating inflation in the general price level. However, the increase in the price of fuel, particularly of petroleum, was greater than that of the general price level. Moreover, since energy is a cost for other industries, the advance in energy prices was a factor in the rising prices of other products. By the same token, the rising prices or costs of non-fuel items used in the energy industry contributed to the increase in energy prices.

But conditions more or less peculiar to the energy industry were largely responsible for the very great increases in the prices of some forms of energy. Shortages and the skyrocketing prices of imports were particularly important in the increase in the prices of oil and gas.

The magnitude of the rise in energy costs differed substantially among U.S. regions in 1973. The variation was due in part to differences in the price increases for various fuels and to regional differences in the degree of dependence on particular fuels. For instance, fuel oil prices showed a much larger increase than the prices of either gas or coal. The Northeast, which is mainly dependent upon fuel oil, experienced a larger increase in energy costs than the Midwest, where gas and coal are relatively more important. Moreover, considerable regional variation was evident in price increases for the same fuels. Areas in which imported oil was more important had larger price increases than those using domestic oil.

In addition to regional variations in the increases of fuel prices, marked variations in price increases developed for the same fuels in the same area, particularly for petroleum fuels. One company would sell its products at prices noticeably different from another's. This situation resulted from shortages and cost pass-throughs permitted under the price controls.

Another phenomenon in 1973 energy prices was the tendency for prices of larger users to increase relatively more than those of smaller users. This was particularly apparent in the regulated rates of gas and electric utilities; but even in the fuel oil markets the prices of transport loads tended to rise more than tankwagon prices.

Chapter Two

Retail Energy Prices

A. RESIDENTIAL GAS AND ELECTRIC BASELOAD

The rates on the non-substitutable electric baseload, which cover such uses as lighting, refrigeration, and small appliances, showed moderate to large increases in 1973, continuing the upward trend evident since 1969. The typical monthly bills for a given amount of use increased in all nine of the standard metropolitan statistical areas (SMSA) examined (one SMSA in each of the Census regions). (See Table 2–1 below.) The largest increase from January 1973 to January 1974 was in the New York SMSA, 45 per cent, which historically has shown the highest monthly bills. Other SMSAs with very large increases ranging from 20 per cent to 26 per cent were Boston, Los Angeles, and Nashville. All but Nashville are served by utilities that depend considerably on heavy fuel oil, which posted extremely large advances in price during 1973. The Chicago and Minneapolis-St. Paul SMSAs in the East and West North Central regions registered increases in the monthly bill of 3 per cent and 4 per cent respectively, the smallest of the SMSAs investigated.

The substitutable electric baseload monthly bills, which include such uses as water heaters, ranges, and clothes dryers, for which either electricity or gas may be used, followed the same pattern in 1973 as that of the electric non-substitutable baseload—except that the increases were substantially larger. In the New York SMSA, the electric substitutable baseload monthly bill in January 1974 was 58 per cent higher than in the same month of the previous year as compared with a 45 per cent gain in the non-substitutable baseload monthly bill. The same comparison in some other SMSAs reveals the following: Nashville—an increase of 56 per cent in the substitutable baseload monthly bill versus a 26 per cent in-

Table 2-1. Residential Monthly Bill for Gas and Electric Baseload Energy Consumption (In Dollars)

	January 1973			November 1973			January 1974		
Census Region—SMSA	*Non-Substitutable Electric*	*Substitutable Electric*	*Substitutable Gas*	*Non-Substitutable Electric*	*Substitutable Electric*	*Substitutable Gas*	*Non-Substitutable Electric*	*Substitutable Electric*	*Substitutable Gas*
New England									
Boston, Mass.	14.50	17.78	14.09	16.23	18.74	15.08	17.73	21.74	15.44
Middle Atlantic									
New York, N.Y.	17.11	21.91	10.44	21.62	28.07	12.24	24.87	34.57	12.33
South Atlantic									
Charlotte, N.C.	11.46	14.06	7.49	12.45	15.89	8.35	12.45	15.89	8.44
East North Central									
Chicago, Ill.	13.79	19.36	6.84	14.42	20.46	7.47	14.39	20.41	7.60
East South Central									
Nashville, Tenn.	8.99	7.95	7.39	9.50	8.78	7.37	11.33	12.43	8.04
West North Central									
Minneapolis-St. Paul	12.64	14.42	7.73	12.80	14.74	8.20	13.01	15.15	8.20
West South Central									
Houston, Tex.	10.58	9.20	6.62	10.98	10.65	6.76	11.07	10.82	6.90
Mountain									
Denver, Colo.	11.26	17.94	4.00	11.85	19.24	—	12.04	19.62	4.74
Pacific									
Los Angeles, Calif.	[a]11.43	14.66	6.96	13.94	18.94	7.36	14.25	19.55	7.36
	[b]12.35	13.44		14.76	18.37		14.76	18.37	

[a]For electric base loads—government utility.

[b]For electric base loads—investor-owned utility.

Note: Non-substitutable base load assumes no electric substitutable base load.
Substitutable base load—electric—assumes no electric space heating.
—gas—assumes no gas space heating
Taxes included where applicable.

Source: American Gas Association *Rate Book;* FPC *National Rate Book;* gas and electric utilities.

crease in the non-substitutable baseload monthly bill; Los Angeles—37 per cent versus 20 per cent (based on the investor-owned utility); and Houston—18 per cent versus 5 per cent. In only one of the nine SMSAs covered in the study, Boston, was the increase in the substitutable baseload monthly bill no higher than that of the non-substitutable baseload monthly bill, with both posting the same relative increase, 22 per cent.

It has been typical for electricity rates to decrease as the volume increases to encourage greater use. Although this situation may still prevail, since the substitutable baseload is additional usage over and above the non-substitutable baseload, rates in 1973 have tended to reduce the advantage of increased use.

The gas substitutable baseload monthly bill rose significantly in SMSAs in most regions of the country, but the increases tended to be less than those of the electric substitutable baseload monthly bills. From January 1973 to January 1974 gas substitutable baseload monthly bills showed increases ranging from a low of 4 per cent in the Houston SMSA to 19 per cent in the Denver SMSA. The largest increases, in excess of 10 per cent, were in the Denver, New York, Charlotte, and Chicago SMSAs; Houston and Minneapolis-St. Paul experienced the smallest. In six of the nine SMSAs the increases in the gas substitutable baseload monthly bill showed a smaller relative increase than the electric substitutable baseload bill. Thus, the gas substitutable monthly bill, which had been lower than that of electricity for the comparable uses constituting this substitutable baseload, tended to show a wider price differential over the electric bill during 1973.

B. RESIDENTIAL HOUSE HEATING

In Most SMSAs the cost of heating a house rose substantially in 1973, irrespective of the form of energy used—gas, oil or, electricity. While the largest increases were in the New England and Middle Atlantic SMSAs of Boston and New York, every one of the SMSAs in every Census region experienced a sharp increase in the cost of house heating in at least one of the energy forms (See Table 2-2 below.)

The most outstanding increases were in fuel oil cost, while the smallest were in gas. Based on January 1974 prices, the average monthly bill for #2 fuel oil for the heating months, assuming normal temperatures, was from 59 per cent to 90 per cent higher in SMSAs where fuel oil is a significant source of home heat[a] than it would have been on the basis of January 1973 prices. The largest increase was in Minneapolis-St. Paul, and the smallest in Boston. Meanwhile, the average monthly gas heating bill, using January 1974 rates, ranged from 6 per cent to 26 per cent

[a] SMSAs with 10 per cent or more of dwellings heated with fuel oil.

Table 2-2. Residential Monthly Bill for Gas, Electric and Fuel Oil Space Heating Use (In Dollars)

Census Region—SMSA	January 1973			November 1973			January 1974		
	Gas	Electric	Oil	Gas	Electric	Oil	Gas	Electric	Oil
New England									
Boston, Mass.	41.72	91.36	34.88	48.61	94.15	47.26	50.63	112.50	55.27
Middle Atlantic									
New York, N.Y.	30.37	87.13	31.08	36.05	107.11	44.22	36.42	142.27	54.98
South Atlantic									
Charlotte, N.C.	18.65	47.57	22.18	20.32	55.48	28.94	20.70	55.48	39.60
East North Central									
Chicago, Ill.	28.59	73.06	32.37	31.47	77.74	41.33	32.30	77.41	54.77
East South Central									
Nashville, Tenn.	26.41	44.78	—	25.97	49.62	—	29.62	71.16	—
West North Central									
Minneapolis-St. Paul	26.26	86.96	35.26	27.84	89.12	48.98	27.84	92.06	67.05
West South Central									
Houston, Tex.	9.11	21.92	—	9.46	25.58	—	9.76	26.01	—
Mountain									
Denver, Colo.	11.42	79.28	—	—	85.92	—	14.37	88.31	—
Pacific									
Los Angeles, Calif.	8.91	36.31[a] 31.92[b]	—	9.86	47.48[a] 45.51[b]	—	9.86	49.12 45.51	—

[a]Government utility.

[b]Investor-owned utility.

Note: 1. For electric space heat assumes an electric substitutable base load and for gas heat assumes a gas substitutable base load.
2. The monthly bills are based on the rates or prices in effect in the designated months but the volume of energy used is the average for the heating months with normal temperatures.
3. Taxes included where applicable.

Source: American Gas Association, *Rate Book;* FPC *National Electric Rate Book;* B.L.S. *Retail Prices and Indexes of Fuels and Electricity;* gas and electric utilities and oil companies.

higher in the various SMSAs than one based on the January 1973 rates. The largest increases in the cost of gas home heating were in Denver, 26 per cent, and Boston, 21 per cent, but seven of the nine SMSAs studied showed increases of 11 per cent or more. Electricity costs for home heating in a majority of the SMSAs posted increases greater than those for gas heating, but less than those for oil heating. The increases ranged from 6 per cent in both Chicago and Minneapolis-St. Paul to 63 per cent in New York. The cost of electric heating continued to be substantially higher than that of either gas or oil in all the SMSAs.

C. INDUSTRIAL USE

Prices of energy for large industrial users in 1973 followed generally the same pattern as for residential users in the various SMSAs, but the industrial increases tended to be relatively larger than the residential. The largest increases were in SMSAs in the Northeast, but some very sharp rises for particular energy sources occurred in SMSAs in other regions. Of the major energy sources of large industrial users, #6 fuel oil registered the highest increases, while the prices of firm gas posted the lowest. The rise in fuel oil prices accelerated dramatically in the last few months of the year.

Electricity rates to large industrial users in the nine SMSAs in the various Census regions rose between 6 per cent and 58 per cent from January 1973 to January 1974. The largest increases were in Nashville and New York. (See Table 2–3 below.) The smallest was in Houston. Four of the SMSAs experienced increases in excess of 40 per cent, while three had increases of 8 per cent or less.

Gas prices to large industrial users showed substantial increases during 1973 in all the SMSAs under consideration, with the rates for interruptible gas generally rising much more than those for firm gas. (See Table 2–4 on the following page.) The rate increases for firm gas ranged from 8 per cent to 30 per cent between January 1973 to January 1974, while those for interruptible gas ranged from 17 per cent to 58 per cent. In fact, interruptible gas in New York was up 243 per cent, but this resulted from the introduction of a new rate schedule rather than from a change in the existing rates. Moreover, in Minneapolis-St. Paul the rate schedules for interruptible gas were eliminated.

The increase in industrial gas prices during 1973 did not follow any regular geographic pattern. For firm gas the smallest increases were in Minneapolis-St. Paul (8 per cent) in the West North Central region, and New York (9 per cent) in the Middle Atlantic region, while the largest were in Denver (30 per cent) in the Mountain region, and Boston (22 per cent) in New England. Meanwhile, for interruptible gas to large

Table 2-3. Electricity—Price to Large Industrial Users (Cents Per Kwh)

Census Region—SMSA	*January 1973*	*November 1973*	*January 1974*
New England			
Boston, Mass.	1.44	1.76	2.14
Middle Atlantic			
New York, N.Y.	2.57	3.24	4.05
South Atlantic			
Charlotte, N.C.	0.83	1.01	1.01
East North Central			
Chicago, Ill.	1.39	1.49	1.48
East South Central			
Nashville, Tenn.	0.77	0.84	1.22
West North Central			
Minneapolis-St. Paul	1.20	1.24	1.29
West South Central			
Houston, Tex.	0.72	0.74	0.76
Mountain			
Denver, Colo.	0.94	1.06	1.11
Pacific			
Los Angeles, Calif.	[a]1.07	1.44	1.51
	[b]1.03	1.57	1.57

[a]For government utility.
[b]For investor-owned utility.
Note: Taxes included where applicable.
Source: FPC *National Electric Rate Book;* electric utilities.

industrial users, Nashville in the East South Central region and Los Angeles on the Pacific Coast had moderate increases of 17 per cent and 19 per cent respectively, while Charlotte in the South Atlantic region and Denver, had increases of 58 per cent and 43 per cent respectively, which were among the highest.

The prices shown for #2 fuel oil and #6 fuel oil on the succeeding two tables are ranges of price postings of some of the various oil companies marketing in the particular SMSA. Prices of fuel oil to large industrial users are generally negotiated and have involved discounts off the posted prices. During the past year of tight supplies and shortages, the discounts have tended to narrow or even be replaced with premiums insofar as price controls permitted. Although ranges in postings have not been uncommon in the past, the ranges have widened greatly during the current period of shortages.

The wide ranges in prices that have developed are a phenomena of shortages and price controls. Under price controls (see Chronology of Price Controls on Petroleum Industry in the Appendix), oil companies have been permitted to pass through higher costs stemming

Table 2-4. Gas—Prices of Firm and Interruptible to Large Industrial Customers (Cents Per Mcf)

Census Region—SMSA	*January 1973*		*November 1973*		*January 1974*	
	Firm	*Interruptible*	*Firm*	*Interruptible*	*Firm*	*Interruptible*
New England						
Boston, Mass.	198.7	109.6	233.3	129.4	242.4	147.3
Middle Atlantic						
New York, N.Y.	148.6	52.2	159.7	144.0[a]	161.5	179.0
South Atlantic						
Charlotte, N.C.	81.0	49.0	89.9	62.5	92.0	77.6
East North Central						
Chicago, Ill.	83.4	49.3	94.6	67.1	98.3	67.5
East South Central						
Nashville, Tenn.	71.7	43.0	68.1	39.1	84.3	50.5
West North Central						
Minneapolis-St. Paul	76.7	—	82.9	—	82.9	—
West South Central						
Houston, Tex.	N.A.	N.A.	N.A.	N.A.	N.A.	N.A.
Mountain						
Denver, Colo.	46.3	27.4	—	—	60.0	39.2
Pacific						
Los Angeles, Calif.	72.5	46.8	81.5	55.8	81.5	55.8

[a]New rate schedule; old rate schedule was abolished August 1973.

Note: Taxes included where applicable.

Source: American Gas Association *Rate Book* and gas utilities.

from their use of foreign crude oil and exempt domestic crude. Since the use of these crudes varies considerably from company to company, considerable differences in costs and prices have resulted.

Prices of #2 and #6 fuel oils to large industrial users in SMSAs under study where the use of these fuels is significant rose dramatically from January 1973 to January 1974, with the largest gains occurring in the last few months of the period. Based on the low of the range, the smallest increase in the price postings of #2 fuel oil was in Charlotte, 40 per cent, while the highest was New York, 107 per cent. (See Table 2-5 below.) Four of the five SMSAs using significant amounts of fuel oil experienced increases of 85 per cent or more. Meanwhile, for #6 fuel oil, the increase in the low posting during the 12-month period ranged from 67 per cent in Boston to 193 per cent in New York. (See Table 2-6 on the next page.)

In Chicago and Minneapolis-St. Paul, January 1974 low postings were more than double the postings in those SMSAs a year earlier. The rise in fuel oil prices during 1973 was considerably greater than the rise for other sources of energy to large indurstrial users.

Coal prices to large industrial users, following the pattern of other energy prices, rose substantially in the 12-month period ending

Table 2-5. No. 2 Fuel Oil—Delivered Prices to Large Industrial Users (Cents Per Gallon)

Census Region—SMSA	*January 1973*	*November 1973*	*January 1974*
New England			
Boston, Mass.	12.6-12.9	16.5-25.6	26.0-38.4
Middle Atlantic			
New York, N.Y.	12.4-12.7	16.6-22.7	25.6-31.1
South Atlantic			
Charlotte, N.C.	12.3-13.2	16.5-18.1	17.2-18.1
East North Central			
Chicago, Ill.	12.3-13.3	17.5-24.9	23.0-36.4
East South Central			
Nashville, Tenn.	—	—	—
West North Central			
Minneapolis-St. Paul	11.5-12.5	15.7-19.1	21.3-31.1
West South Central			
Houston, Tex.	—	—	—
Mountain			
Denver, Colo.	—	—	—
Pacific			
Los Angeles, Calif.	—	—	—

Note: Taxes included where applicable.

Source: Prices from Platt's *Oilgram Price Service;* taxes from Commerce Clearing House; delivery charges estimated.

Table 2-6. No. 6 Fuel Oil—Delivered Prices to Large Industrial Users (Dollars Per Barrel)

Census Region—SMSA	*January 1973*	*November 1973*	*January 1974*
New England			
Boston, Mass.[a]	$5.22	$5.71-8.47	$8.70-12.46
Middle Atlantic			
New York, N.Y.[b]	5.33	8.81	15.61
South Atlantic			
Charlotte, N.C.	—	—	—
East North Central			
Chicago, Ill.[c]	5.17-5.32	7.36	11.33-11.77
East South Central			
Nashville, Tenn.	—	—	—
West North Central			
Minneapolis-St. Paul[d]	4.58	5.85-7.22	9.64-10.37
West South Central			
Houston, Tex.	—	—	—
Mountain			
Denver, Colo.	—	—	—
Pacific			
Los Angeles, Calif.	—	—	—

[a]0.5% sulfur maximum.
[b]0.3% sulfur maximum.
[c]1.0% sulfur maximum.
[d]Regular.

Note: Taxes included where applicable.

Source: Prices from Platt's *Oilgram Price Service;* taxes from Commerce Clearing House; delivery charges estimated.

January 1974. (See Table 2-7 on the next page.) Four of the five SMSAs where coal is an appreciable source of energy for industry showed increases in industrial coal prices of 15 per cent or more. The largest increase occurred in Chicago, 22 per cent. Minneapolis-St. Paul, however, was an exception, with only a small increase of 2 per cent. In the SMSAs where industrial use of coal is important, the rise in coal prices tended to be less than the rise for other sources of energy used by large industrial users.

The coal prices used in this study were contract prices, the most typical for the coal used by large industrial users. Nevertheless, insofar as the industrial users made spot purchases of coal, a substantially different picture would emerge. Spot coal prices have shown considerably larger increases than contract prices, with spot coal prices doubling during the past year in some areas.

D. MOTOR GASOLINE

Service station prices of major brand regular grade gasoline rose markedly during 1973 in all the SMSAs after following an irregular but moderately

Table 2-7. Coal—Delivered Prices to Large Industrial Customers (Dollars Per Ton)

Census Region—SMSA	*January 1973*	*November 1973*	*January 1974*
New England			
Boston, Mass.	—	—	—
Middle Atlantic			
New York, N.Y.	—	—	—
South Atlantic			
Charlotte, N.C.	20.59	22.28	23.69
East North Central			
Chicago, Ill.	13.68	16.04	16.75
East South Central			
Nashville, Tenn.	21.15	23.12	24.25
West North Central			
Minneapolis-St. Paul	28.50	28.80	29.09
West South Central			
Houston, Tex.	—	—	—
Mountain			
Denver, Colo.	8.31	9.58	9.80
Pacific			
Los Angeles, Calif.	—	—	—

Note: Includes taxes where applicable.
Source: Major coal companies.

upward trend since 1965. (See Table 2-8 on the next page.) The largest increases in service station prices in the twelve months ending in early January 1974 were in the Northeastern section of the Nation, with prices in Boston and in New York up 25 per cent and 26 per cent respectively. The smallest relative increases were in Charlotte and Chicago, both showing 10 per cent increases.

There are a number of aspects of the retail gasoline market that must be considered in order to understand price developments. In the first place, marked variations have existed in recent months in the prices of the major brand service stations in the same SMSA. Shortages and price controls permitting a cost pass-through on foreign and exempt domestic crude or of imported refinery products promote these variations. Therefore, while the prices shown on Table 2-8 are believed to be reasonably representative, some wide deviations from these prices existed at a substantial number of branded stations in some SMSAs in January 1974.

At the time of the January 1973 prices, depressed pricing still prevailed in some areas of the Nation. In January 1974 practically all service station prices were at price ceiling levels. Consequently, the large increases from January 1973 to January 1974 in some SMSAs—Denver and Boston for example—reflect not only the rising costs that have affected gasoline prices in all markets but also the elimination of the element of depressed pricing.

Table 2-8. Service Station Prices of Major Brand Regular Grade Gasoline (Cents Per Gallon)

Census Region—SMSA	January 1973			November 1973			January 1974		
	Dealer TW Price (Excl. Tax)	Tax	Service Station Price (Incl. Tax)	Dealer TW Price (Excl. Tax)	Tax	Service Station Price (Incl. Tax)	Dealer TW Price (Excl. Tax)	Tax	Service Station Price (Incl. Tax)
New England									
Boston, Mass.	18.0	11.5	35.9	20.7	11.5	39.9	25.3	11.5	44.9
Middle Atlantic									
New York, N.Y.	20.1	15.2	41.9	22.4	15.2	44.9	27.0	15.2	52.9
South Atlantic									
Charlotte, N.C.	18.8	13.0	38.9	19.8	13.0	39.9	21.8	13.0	42.9
East North Central									
Chicago, Ill.	19.6	13.0	40.4	21.3	12.9	42.1	23.8	13.2	44.6
East South Central									
Nashville, Tenn.	17.8	11.0	35.9	20.1	11.0	38.9	22.1	11.0	40.9
West North Central									
Minneapolis-St. Paul	18.2	11.0	36.9	21.3	11.0	40.6	23.8	11.0	43.1
West South Central									
Houston, Tex.	16.3	9.0	31.9	18.7	9.0	34.9	21.1	9.0	37.3
Mountain									
Denver, Colo.	16.8	11.0	34.9	21.3	11.0	40.6	23.8	11.0	43.1
Pacific									
Los Angeles, Calif.	17.9	11.6	37.9	21.3	11.6	40.0	23.9	11.6	44.7

Note: (1) Prices for January and November 1973 are as of the first day of the month; prices for January 1974 are as of one day within the first week of that month.

(2) Dealer tank wagon prices are the prices the service station dealers pay their suppliers.

Source: Platt's *Oilgram Price Service;* oil companies.

The relative increases in gasoline prices during the past year have been considerably less than those for the petroleum products #2 and #6 fuel oils. In part, this is because the proportion of taxes included in gasoline prices is much heavier than in fuel oil prices and because taxes remained unchanged or rose less than the cost of the gasoline. Another factor is that price controls permitted a greater upward adjustment in fuel oil prices than in gasoline prices in order to encourage an increase in fuel oil supplies to meet winter heating needs in 1974. Also, imports supply most of the heavy fuel used in this country and more distillate fuel oil than gasoline. Delivered prices of imported products have been considerable higher than those of domestic products.

Chapter Three

Primary Prices of Fossil Fuels

The large increases in energy prices during 1973 extend all the way from the retail level back to the primary sources of the energy. While it is not within the scope of this study to determine the extent of the price or cost increases at the various levels of production and distribution, an examination of price developments at the primary level should be helpful in understanding the rise in energy prices at the retail level. Accordingly, a brief review is included of the price of some representative domestic crude oils, the price of one of the key foreign crude oils, the field price of interstate gas and the Canadian border prices of gas, and the mine price of coal.

A. CRUDE OIL

The cost of crude oil to the nation's refineries rose sharply in 1973, with the largest increases occurring toward the end of the year. The estimated price of some representative domestic crudes subject to price controls and delivered to four major refinery centers of the Nation—Philadelphia, Houston, Chicago, and Los Angeles—were from 44 per cent to 49 per cent higher in January 1974 than a year earlier. (See Table 3–1 on the next page.) If domestic crude exempt from price controls[1] had been used, the increases would have been higher.

Meanwhile, foreign crude prices showed exceptionally large increases during 1973, far exceeding those of crude produced in the United States. The price of Arabian light crude;[a] one of the key foreign

[a]See Appendix, Chronology of Price Controls on Petroleum Industry, for a definition of domestic crude oil exempt from price controls.

[a]Arabian light crude is not the most commonly used foreign crude in United States refineries, but its price movements tend to be representative of those of other foreign crudes.

Table 3-1. Estimated Crude Oil Prices[1] Delivered to Selected Refinery Centers (Dollars Per Barrel)

	January 1973	*October 1973*	*January 1974*
Philadelphia			
South Louisiana	4.29	5.05	6.09
West Texas Sweet	4.19	5.10	6.14
West Texas Sour	4.08	4.99	6.03
Arabian Light	3.10	5.50	10.36
Houston			
South Louisiana	3.80	4.40	5.40
West Texas Sweet	3.69	4.44	5.44
West Texas Sour	3.58	4.33	5.33
Arabian Light	3.25	5.60	10.45
Chicago			
South Louisiana	3.96	4.56	5.56
West Texas Sweet	3.80	4.55	5.55
West Texas Sour	3.69	4.44	5.44
Arabian Light	3.46	5.79	10.65
Los Angeles			
L. Beach (Signal Hill) Huntington Beach	3.82	4.42	5.42
Arabian Light	3.16	5.45	10.30

[a]All crude oils are 36° gravity.

Source: See Tables 1, 2, 3, 4, and 5 in Appendix.

crudes, delivered to the Philadelphia refineries, more than tripled in the twelve months ending January 1974. The crude price used in this study was for equity crude owned by the oil company producers. Equity crude constituted the major portion of foreign crude produced in 1973. Even higher prices were obtained for crudes owned by the foreign countries and sold or auctioned in the latter part of 1973.

Increases in the delivered prices of foreign crude as well as of domestic crude used in the Philadelphia refineries were inflated by appreciable increases in tanker rates. (See Tables 1, 2, 3, 4, and 5 in the Appendix.) Nevertheless, most of the increases in the delivered prices stemmed from the increases in the prices of the crude oil itself, the field price of United States crude and the tax-paid cost[b] of foreign crude.

The field price of domestic crude, after a period of stable prices in 1971 and 1972, began to rise in March 1973 by some 25 cents per barrel. (See Table 3-2 on the next page.) In the succeeding few months the increase broadened and accelerated until the Cost of Living Council on June 13, 1973 instituted a freeze on prices of refined products and crude oils. The freeze was lifted on August 19 as Phase IV price controls were introduced.

[b]Tax-paid cost consists of the royalty and taxes on the crude paid by the producing company to the host foreign country and the production costs on the crude.

Phase IV regulations permitted an increase of 35 cents per barrel of crude over the highest price prevailing in a field on May 15, 1973. Moreover, new crude oil (defined as any crude produced and sold from a field in excess of the 1972 level), together with an equivalent amount of old crude from the field, was exempt from price controls. Then on November 16, 1973, crude oil produced from stripper wells (defined as the wells producing less than 10 barrels per day) also was made exempt by law from price controls. On December 19, the Cost of Living Council permitted a $1.00 per barrel increase in the existing ceiling prices on domestic crude oil.

Table 3-2. Posted Prices[1] of Selected 36° Gravity Domestic Crude Oil (Dollars Per Barrel)

Year	*South Louisiana*	*West Texas Sweet*	*West Texas Sour*	*L. Beach (Signal Hill) Huntington Beach California*
1960	3.02	2.83	2.79	3.27
1965	3.11	2.93	2.83	3.27
1966	3.11	3.00	2.87	3.27
1967	3.11	2.98	2.92	3.27
1968	3.18	3.08	2.92	3.27
1969	3.38	3.23	3.06	3.47
1970	3.63	3.48	3.31	3.72
1971	3.63	3.48	3.37	3.72
1972	3.67	3.48	3.37	3.72
1973 June 15[b]	3.92	3.73	3.54	3.97
1973 Oct. 15[b]	4.27	4.23	4.12	4.32
1974 Jan. 15[b]	5.27	5.23	5.12	5.32

[a]Lowest posted price at end of years 1960-1972, exclusive of local port or other governmental charges, sales taxes, etc.

[b]Prices are for crude oil under price controls, which comprises the major portion of domestic crude production. Prices of crude oil exempt from price ceilings on January 15, 1974 were reported as high as $10.40 in the Texas Coast area.

Source: Platt's *Oil Price Handbook* and *Oilgram Price Service*.

The prices of foreign crude, which had been cheaper than domestic crude during most of the post-World War II period, in 1973 moved to levels substantially higher than domestic crude. The Mandatory Oil Import Program and recent price controls have been factors tending to encourage price differentials between domestic and foreign crude oil.

The sharp increases in foreign crude oil prices in 1973 represented a continuation of an upward trend that started in the latter part of 1970. (See Table 3-3 on posted prices on the next page.) This upward trend has been associated with actions taken by the OPEC countries (Organization of Petroleum Exporting Countries), which possess the bulk of the free world's oil reserves. (See OPEC Chronology in the Appendix.) However, other factors have contributed, including: (1) the rapid growth in the world's oil demand, which has exceeded expectations in recent years, and (2) a tight tanker situation with high tanker rates.

Table 3–3. Posted Prices Arabian Light 34° Gravity Crude Oil FOB RAS Tanura (Dollars Per Barrel)

December 31, 1960	1.80
December 31, 1965	1.80
December 31, 1970	1.80
February 15, 1971	2.18
June 1, 1971	2.29
January 20, 1972	2.48
January 1, 1973	2.59
April 1, 1973	2.74
June 1, 1973	2.90
July 1, 1973	2.96
August 1, 1973	3.07
October 1, 1973	3.01
October 16, 1973	5.12
November 1, 1973	5.18
December 1, 1973	5.04
January 1, 1974	11.65

Source: Platt's *Oil Price Handbook;* U.S. Department of Interior, Office of Oil and Gas, *Third Report—World Wide Crude Oil Prices—Summer 1973; Petroleum Intelligence Weekly.*

B. NATURAL GAS

The average field price of natural gas sold to interstate pipeline companies in 1973 registered a pronounced increase, continuing the rising trend evident during the 1960's and early 1970's. (See Table 3–4.) In November 1973, the latest month for which data are available at the time of this writing, the average field price to interstate pipelines was 23.85 cents per Mcf (thousand cubic feet), which was about 2.5 cents per Mcf or 12 per cent higher than in January 1973.

Canada has been an important source of gas for the Pacific Coast and the upper Midwest. The border price of Canadian gas showed an appreciably larger absolute increase in 1973 than the field price of interstate gas in the United States, although the relative increase was the same. The border price of Canadian gas in November 1973 averaged 35.17 cents per Mcf, up 3.74 cents per Mcf or 12 per cent over the January 1973 level. The border price of Canadian gas has been rising since 1968, and the extent of both the relative and the absolute increase since that time has been considerably larger than that of the field price of interstate gas in the United States.

Table 3-4. Wholesale Prices of Natural Gas (¢/Mcf)

Year	*Average Field Price*[a]	*Border Price*[b]	*Average Price For Resale*[c]
1960	15.53	22.90	n.a.
1961	16.30	25.00	n.a.
1962	16.54	24.98	n.a.
1963	16.63	25.06	n.a.
1964	16.55	24.43	n.a.
1965	16.66	24.21	n.a.
1966	16.70	23.96	n.a.
1967	16.96	24.12	35.28
1968	17.20	23.87	35.17
1969	17.48	24.57	36.09
1970	17.93	25.79	38.07
1971	19.13	28.00	42.23
1972	20.54	30.61	45.87
Jan. 1973	21.37	31.43	46.40
Nov. 1973	23.85	35.17	51.82

[a]Average field price of natural gas sold to interstate pipelines.
[b]Average Canadian border price of natural gas sold to major pipelines.
[c]Average price of natural gas sold by interstate pipelines for resale. 1969, 1970, Nov. 1973, and Jan. 1973 prices are averages of 33 companies. 1967, 1968, 1971 and 1972 prices are averages of 30 companies.

n.a.—not available.

Source: U.S. Federal Power Commission—*Sales by Producers of Natural Gas to Natural Gas Pipeline Companies, Sales by Producers of Natural Gas to Interstate Pipelines Companies, Average Price Received by Major Pipelines.* and news releases.

The increases in the field price and the Canadian border price to interstate pipelines has been translated into increases in the prices these pipelines charge the gas utility distributors, the so-called "city gate" prices. The average price of gas sold by interstate pipelines for resale rose from 46.40 cents per Mcf in January to 51.82 cents per Mcf in November 1973. This increase of 12 per cent was the same relative increase shown by the field price and the Canadian border price.

C. COAL

Bituminous coal prices also registered a marked rise in 1973, thus following the pattern of the other major fossil fuels, oil and gas. The mine price of bituminous coal in industrial size screenings increased 17 per cent[a] between January 1973 and December 1973. (See Table 3–5 on the next page.) This increase was a continuation of the upward trend in coal prices that prevailed during the last half of the 1960's and in the early 1970's.

aThis percentage increase was derived from U.S. Bureau of Labor Statistics data, which have not included the prices of Western coal.

Table 3-5. Mine Price of Bituminous Coal Industrial Size-Screening United States[a] (Dollars Per Ton)

January 1960	5.19
January 1961	5.15
January 1962	5.02
January 1963	4.74
January 1964	4.73
January 1965	4.79
January 1966	4.79
January 1967	5.12
January 1968	5.28
January 1969	5.80
January 1970	6.53
January 1971	9.75
January 1972	10.27[e]
January 1973	11.21
December 1973	12.10

[a]Prices are limited to Eastern coal.

[e]Estimated.

Source: U.S. Bureau of Labor Statistics, *Wholesale Prices and Price Indexes.*

Important factors in the rise in coal prices in 1973 were (1) the continuing increases in production costs associated with the impact of inflation on the cost of materials and (2) higher labor costs—both wage and non-wage labor costs, such as welfare payments. Also contributing to the rise in coal prices were increases in the demand for coal, particularly following the Arab oil embargo at mid-October 1973. East Coast utilities capable of reconverting from fuel oil to coal were encouraged to seek new coal supplies. This development occurred at a time when the coal industry had little spare capacity.

Appendix

A. Chronology of Price Controls on the Petroleum Industry
B. Chronology of Price Controls on Gas and Electric Public Utilities
C. Chronology of Price Controls on the Coal Industry
D. OPEC Chronology
E. Tables—Estimated Cost of Representative Crude Oils Delivered to Selected Refining Centers

Table 1—South Louisiana 36° Crude Oil
Table 2—West Texas 36° Sweet Crude Oil
Table 3—West Texas 36° Sour Crude Oil
Table 4—Long Beach (Signal Hill) Huntington Beach 36° Crude Oil
Table 5—Saudi Arabian 36° Light Crude Oil

CHRONOLOGY OF PRICE CONTROLS ON THE PETROLEUM INDUSTRY

Phase I

August 15, 1971—Price controls on the petroleum industry begin on this date with the institution of Phase I. Guidelines for the petroleum industry are included in those set for the economy in general. All prices, rents, wages, and salaries are to be temporarily frozen for a 90-day period beginning on this date. Prices, rents, wages, and salaries are set at levels not greater than the highest of those for a substantial volume of actual transactions by each individual, business, firm, or other entity of any kind during the 30-day period ending August 14, 1971. This temporary freeze is designed to provide the necessary time to develop a plan, called Phase II, for a more flexible stabilization program.

Also established under Phase I is the Cost of Living Council, which is to be composed of the heads of several government agencies. The Council will be the main force governing further price control programs.

Phase II

November 15, 1971—Phase II controls are put into effect after the 90—day price freeze. Once again standards set for the petroleum industry are included in those set for the remainder of the economy.

Two new stabilization agencies are created under Phase II: the Pay Board and the Price Commission. The Pay Board, composed of five representatives of labor, five representatives of management, and five public members, is responsible for developing specific standards and criteria for pay adjustments consistent with the goals of the stabilization program. The Price Commission, consisting of seven public members, is established to develop standards and criteria for adjustments in prices and rents.

Wages, under the guidelines of the Pay Board, are to be limited to increases not in excess of 5.5 per cent. Meanwhile, price ceilings are set for petroleum refiners (and other manufacturing establishments) by the Price Commission at the level of the base price in the base period plus increases in cost so long as profit margins do not exceed the best two of three fiscal years prior to August 15, 1971. The base period is the thirty days prior to August 15, 1971, and the base price is the highest price at which 10 per cent of transactions were made with a given class of customers during the base period. Wholesalers and retailers are to limit their markups to those in effect during the base period.

All enterprises are placed into one of three categories according to their volume of sales. Tier I firms are those with gross sales of $100 million or more, and they must have the approval of the Price Commission before any price increases may be put into effect. Tier I firms must submit quarterly price, cost, and margin reports to the Price Commission. Tier II firms, those with gross sales between $50 million and $100 million, are not required to secure Price Commission approval for price increases, but must submit quarterly price, cost, and margin reports to the Commission, just as Tier I firms do. Tier III firms, those having less than $50 million gross sales, are not required to notify the Price Commission in advance of price increases; nor are they required to submit quarterly reports. Nevertheless, the Tier III firms are governed by the same standards in pricing as the larger firms and are subject to monitoring and spot checks.

Phase III

January 11, 1973—Phase III regulations begin on this date and may be characterized as a general loosening of the strict Phase II program. The Price Commission and the Pay Board are abolished and their responsibilities placed under the Cost of Living Council. Again, regulations affecting the petroleum industry follow the general provisions set by the Cost of Living Council. The general price standard set is that prices may be increased over those authorized on January 10, 1973 so long as the profit margin does not increase. Prices may be increased, however, by a weighted annual average of 1.5 per cent over those permitted on January 10, 1973 to reflect increased costs without regard to profit margins. Prenotification of price changes is no longer required under Phase III. Controls are to be applied voluntarily on a self-administered basis. Quarterly reports on prices, costs, and margins are required from those firms with annual sales over $250 million.

March 6, 1973—Mandatory price controls are imposed by the Cost of Living Council on the petroleum industry. These new regulations replace the voluntary, self-administered rules. The mandatory controls apply to firms that derive $250 million or more of their revenues from the sale of covered products. The covered products include petroleum products, either refined or purchased for resale; and crude oil, either produced or acquired domestically or imported for resale.

The mandatory controls apply to the sale of covered items at all levels—refining, wholesale, and retail. The base price is set at the price authorized on January 10, 1973. Price increases without cost justification are limited to a weighted annual average of 1 per cent above the base price. Increases above that figure up to 1.5 per cent are subject to a cost justification. Advance notice of any increase over 1.5 per cent must be given to the Cost of Living Council; the increase is subject to profit margin limitations as had been the case during Phase II controls. Quarterly reports of prices and margins must be submitted.

Firms with annual sales of under $250 million are not affected by the mandatory controls. These smaller firms remain under the control of the original Phase III guidelines. Controls remain voluntary and self-administered. No prenotification or quarterly reports are required.

May 11, 1973—The base price effective under mandatory controls is revised as of this date. The purpose of this ruling is to facilitate crude oil imports to increase supply. The base price previously was set at the price authorized on January 10, 1973. The base price now may change each time the cost of the seller increases. The seller may now pass through costs over which he has no control.

June 13, 1973—The President announces a freeze for not more than sixty days on all prices. Prices and profit margins are frozen at the highest level at which a substantial number of transactions occurred during the period June 1, 1973 through June 8, 1973. The purpose of the freeze is to halt the rapid price increases resulting from the loosened Phase III controls.

August 10, 1973—The Cost of Living Council announces its decision to extend the freeze on prices charged for crude petroleum and refined products until August 19, 1973. The purpose of the limited extension is to adequately consider all comments received on the proposed Phase IV regulations concerning petroleum.

Phase IV

August 17, 1973—Phase IV oil regulations are put into effect on this date. All firms in the petroleum industry regardless of size are governed by these rules. Quarterly reports on prices, cost, and margins are required for firms with over $50 million in annual sales. Firms with sales under $50 million are subject to spot checks and must submit reports on request.

Producers have ceiling prices imposed on domestic crude oil. Prices are fixed at the highest posted price on May 15, 1973, plus an amount not in excess of 35 cents per barrel. To encourage increased production of domestic crude, a two-tiered price system is established. Prices of "old" domestice crude are subject to the aforementioned ceiling price. Any "new" crude and an equivalent amount of "old crude is not subject to the ceiling price or otherwise controlled. "New" crude is defined as the crude produced from a given field in a given month in excess of the crude produced in the same field in the same month of 1972.

Refiners' base prices for petroleum products are established at May 15 levels. Increases are allowed only on a dollar-for-dollar basis to recover increased raw material costs. Rules apply to all sales by refiners to wholesalers, jobbers, other resellers and retailers of their product.

Retail sales of gasoline, #2 heating oil, and #2 diesel fuel are also placed under controls. The ceiling prices for #2 heating oil become effective August 19, while the ceiling prices for gasoline and #2 diesel fuel become effective August 31, 1973. In all cases, the ceiling price is defined as the average cost of inventory on August 1, 1973, plus the average markup as of January 10, 1973. In computing the ceiling prices of gasoline and #2 heating oil, retailers may use an actual markup of seven cents per gallon. This is to avoid any undue hardship for those retailers who were engaged in price wars on January 10, 1973. Monthly increases in the ceiling prices of #2 heating oil beginning September 1, 1973 are permitted to reflect increased cost of importing the product.

Effective September 1, 1973, retailers of gasoline and #2 diesel fuel must post the minimum octane number for sales of gasoline. Octane number is defined as the sum of the Research and Motor octane numbers divided by two.

Finally, a base rent is established with respect to a lease on real property used in the retailing of gasoline. No increases in rent are permitted above the amount charged for that station pursuant to contractual terms prevailing on May 15, 1973.

September 28, 1973—It is announced on this date that retailers of gasoline, diesel fuel, and home heating oil may increase the ceiling price. The new ceiling price is defined as the actual May 15, 1973 selling price plus increased costs of imports, domestic crude, and purchased product from May 1973 to September 1973.

October 15, 1973—A series of modifications is released dealing with the ceiling prices of gasoline, home heating oil, and diesel fuel. An upward ceiling adjustment is authorized immediately to reflect increased product costs incurred up to October 15 since the last ceiling price adjustment at the end of September. No additional price increases will be allowed at any level of distribution for the three products between October 15 and October 31. The two-week freeze will be used to evaluate comments received on proposed regulations.

October 31, 1973—The Cost of Living Council announces changes in Phase IV price controls on diesel fuel, gasoline, and home heating oil. The new ruling calls for a dollar-for-dollar pass-through of increased raw material costs of these products at all levels of distribution on a once-monthly basis. The amendment also requires all retailers of gasoline and diesel fuel to post new ceiling price stickers on each pump by November 21.

November 16, 1973—A new law becomes effective governing the pricing of certain crude oils in the United States. The law exempts the prices charged on the first sale of crude petroleum produced from stripper wells, i.e., wells that produced an average of ten barrels or less per day during the preceding calendar month. A well must have been operating at the maximum feasible rate of production in accord with conservation practices, and its output must not have been curtailed by mechanical disruption, in order to qualify.

December 3, 1973—The Cost of Living Council issues amendments concerning the allocation of increased product costs by petroleum refiners. Regulations are effective as of November 30, 1973 and are to be used in computing December price adjustments. The regulation requires refiners to include all product cost increases in the formula for allocation of increased product costs. Previously, refiners reselling imported product could pass through increased costs of high-cost imported product to independent retailers, while at the same time selling their own domestically refined products through branded dealers at lower prices.

December 5, 1973—The first step of the Refinery Balance Incentive Program is put into effect to encourage the production of distillate fuel. Refiners are allowed to increase the price of distillates two cents per gallon and are required to reduce gasoline prices by one cent per gallon. The adjustments are to be reflected in the January product prices. Under this plan, distillate fuels are defined to be #1 and #2 home heating oils, #1 and #2 diesel fuel, #4 distillate, #4 diesel fuel, kerosene, and naphtha- and kerosene-base jet fuels.

December 19, 1973—The Cost of Living Council permits an increase in the ceiling price of domestic crude oil. This one-time increase is to be passed along on a once-a-month basis and is set at $1.00 per barrel. The new ceiling price now stands at the May 15, 1973 posting, plus a maximum of $1.35 per barrel.

December 26, 1973—A change is made in the governing authority over petroleum regulations. The Cost of Living Council hereby delegates to the new Federal Energy Office (FEO) the authority to implement the allocation and price stabilization activities as they relate to petroleum products and crude oil. Any regulation issued prior to this date shall remain in effect unless altered by the new office.

December 28, 1973—The second step of the Refinery Balance Incentive Program is implemented by the FEO. The new ruling extends the refiners' incentive to increase production of distillate fuels. A sliding scale matrix is introduced that permits measured increments in distillate fuel prices to the extent that production of these fuels exceeds the production level resulting from the first step adjustment.

January 15, 1974—Mandatory petroleum price regulations are issued by the FEO. These regulations basically continue previous rules put into effect by the Cost of Living Council. The regulations are now reissued under the authority of the FEO.

For crude petroleum, the ceiling price is set at the highest May 15, 1973 posted price for the given grade of crude, plus a maximum of $1.35 per barrel. For the sale of crude covered under the mandatory allocation program, the price is set at a weighted average crude price, plus a transportation adjustment, plus a 6 per cent handling fee. In addition, for each barrel of crude he sells, the refiner may increase his product prices 84¢ per barrel. Under the mandatory allocation program, crude-long refiners must supply crude-short refiners with oil to run their refineries at a national average level established by the FEO. Also, the law exempting stripper well production from controls remains intact and unaffected by these regulations. "New" crude as previously defined also remains exempt.

Refiners may charge a price in excess of the base price only to recover, on a dollar-for-dollar basis, net increases in allowable costs since the base period. The general rule governing the base price for sales sets the price at the weighted average lawful price in transactions on May 15. The base price for #1 heating oil, #4 fuel oil, #1 and #4 diesel fuel, kerosene and naphtha- and kerosene-base jet fuels may be revised upward to 2 cents over the May 15, 1973 base. The 2 cent increase also applies to #2 diesel fuel and #2 home heating oil. The refiners incentive program in effect previously is to be continued.

Resellers and retailers may not charge on any petroleum products a price exceeding the weighted average price in effect on May 15, 1973, plus an amount reflecting increased costs on a dollar-for-dollar basis. These increases are to be passed along on a once-a-month basis.

February 5, 1974—The FEO issues a special order setting new price guidelines for the sale of #2 diesel fuel at all transaction levels. The new ceiling price is defined as the highest price charged for 10 per cent of the transactions made with a given class of purchaser between February 1, 1974 and February 4, 1974. The order is to be effective for one month or until Congress affords relief to truckers in the form of higher rates.

February 27, 1974—The FEO amends the prevailing regulations covering gasoline and distillate fuels. Priority is now given to the production of more gasoline to meet the rising demand of the coming spring and summer months. Thus, the FEO removes the distillate incentives previously adopted in the first and second steps of the refinery incentive program. Effective March 1, 1974, this will result in a decrease in the May 15, 1973 base selling prices of distillate by two cents per gallon and an increase in the May 15, 1973 base selling prices of gasoline by one cent per gallon.

February 28, 1974—An FEO announcement allows for further increases in the retail prices of gasoline. This new ruling is aimed at granting relief to retail dealers whose margins have declined due to shortage of products. Effective March 1, 1974, retailers receiving less than 85 per cent of their 1972 base period volumes are allowed to increase by one cent per gallon the price of gasoline.

CHRONOLOGY OF PRICE CONTROLS ON GAS AND ELECTRIC PUBLIC UTILITIES

Phase I

August 15, 1971—Price controls on gas and electric public utilities begin on this date with the institution of Phase I. Guidelines for public utilities are included in those set for the economy in general.

All prices, rents, wages, and salaries are to be temporarily frozen for a 90-day period beginning August 15, 1971. Prices, rents, wages, and salaries are set at levels not greater than the highest of those for a substantial volume of actual transactions by each individual, business, firm, or other entity of any kind during the 30-day period ending August 14, 1971. This temporary freeze is designed to provide the necessary time to develop a plan, called Phase II, for a more flexible stabilization program.

Also established under Phase I is the Cost of Living Council, which is to be composed of the heads of several government agencies and will be the main force governing further price control programs.

Phase II

November 14, 1971—Phase II controls are put into effect following the expiration of the 90-day price freeze.

Two new stabilization agencies are created under Phase II, the Pay Board and the Price Commission. The Pay Board is composed of five labor representatives, five management representatives, and five public members and is responsibile for developing specific standards and criteria for pay adjustments consistent with the goals of the stabilization program. The Price Commission, consisting of seven public members, is established to develop standards and criteria for adjustments in prices and rents.

New regulations are established governing the control over rates of public utilities. The new regulations provide that a regulated public utility that had gross receipts of $100 million or more during its most recent fiscal year must inform the Price Commission of all requests for rate increases. The Price Commission reserves the right to review and limit the amount of any such requested increase. The Price Commission also reserves the right to review and limit the amount of any increase approved by a regulatory agency or other appropriate legal authority for regulated public utilities that had revenues between $50 and $100 million during its most recent fiscal year.

January 17, 1972—A new amendment is issued governing price increases by public utilities. The new regulations cover a period of ninety days following this date. Reporting requirements for regulated public utilities will apply only to firms that must give prior notification of price increases and to those whose price increases would raise a utility's aggregate annual revenue by more than 1 per cent. Reporting requirements for unregulated public utilities will apply only to firms that must give notice and to those whose price increases would increase their aggregate annual revenues by 2.5 per cent.

The general rules governing price increases are:

1. The rate increase is cost-based and does not reflect future inflation expectations;
2. The increase is needed to assure continued, adequate, and safe service, or to provide for necessary expansion;
3. The increase will achieve the minimum rate of return needed to attract capital at reasonable rates and not impair the credit of the utility;
4. The increase has been certified by the proper regulatory agency;
5. The increase is consistent with the goals of the Price Commission.

The Price Commission is given ten days in which to act upon proposed increases. If after ten days no action is taken, the increase becomes effective.

Also included in the amendment is a provision requesting regulatory agencies to submit proposals describing types of increases they believe should not be reported due to the insignificance of their inflationary effects. Price Commission approval of such a proposal exempts the reporting of such increases. Proposals are to be received within ninety days from this date.

March 17, 1972—A new set of regulations governing price increases by public utilities is issued. Each federal or state regulatory body is to submit to the Price Commission a proposed set of rules for its use in considering price increases for utilities under its jurisdiction. If the Price Commission approves the rules and accompanying procedures, or fails to act upon them within thirty working days, the regulatory body will be given a certificate of compliance with the Economic Stabilization Program. Upon the agency's adoption of these rules, any price increase granted by the regulatory body is to be considered in compliance with the program and will not be subject to further review. The general criteria governing proposed increases are to center around rates of return. The projected rate of return after the rate increase is not to exceed the rate of return granted to the utility by the last decision of the regulatory agency pertaining to that utility. Finally, under this new amendment all unregulated public utilities and those substantial parts of regulated utilities that are unregulated will, after March 24, 1972, be governed by the Price Commission's regulations pertaining to service organizations.

Phase III

January 10, 1973—Upon the imposition of Phase III regulations, special rules applicable to public utilities are issued. The new standards apply to each legal rate increase effected after January 10, 1973 by any publicly, privately, cooperatively, or municipally owned public utility, whether or not the increase is approved by the regulatory agency. Standards set will apply to both regulated and unregulated utilities.

Increases are to be consistent with these criteria:

1. The increase is cost-justified and does not reflect future inflationary expectations;
2. The increase is the minimum required to assure continued adequate and safe service or to provide for necessary expansion to meet future requirements;
3. The increase will achieve the minimum rate of return needed to attract capital at reasonable costs and will not impair the credit of the public utility; and
4. The increase takes into account expected and obtainable productivity gains.

Also under Phase III, the Pay Board and the Price Commission are abolished. The responsibilities of these two agencies now pass to the Cost of Living Council.

Phase IV

August 17, 1973—Initiation of Phase IV regulations takes place on this date. Under 150.56 of these rules, rate increases for commodities or services provided by a public utility are exempted from CLC control.

November 6, 1973—A clarification is published of those industries classified as exempt due to their status as public utilities. Activities under the following industry group numbers, as appearing in Division E of the Standard Industrial Classification Manual, are exempt: #491, Electric Service; #492, Gas Production and Distribution; and #493, Combination Electric and Gas, and Other Utility Services.

CHRONOLOGY OF PRICE CONTROLS ON THE COAL INDUSTRY

Phase I

August 15, 1971—Price controls on the coal industry begin on this date with the institution of Phase I. Coal guidelines are included in the general economic standards set. All prices, wages, rents, and salaries are to be frozen for a 90-day period beginning on this date. Prices, wages, rents, and salaries are set at a ceiling level not greater than the highest of those pertaining to a substantial volume of actual business transactions by each individual, firm, business, or other entity of any kind during the 30-day period ending August 14, 1971.

The Cost of Living Council is also established under Phase I. The Council is to be composed of the heads of several federal agencies. It will be the main force governing further price control programs.

Phase II

November 15, 1971—Phase II controls are put into effect after the 90-day price freeze. Once again standards set for the coal industry are included in those issued for the remainder of the economy.

Two new stabilization agencies are created under Phase II: The Pay Board and the Price Commission. The Pay Board, composed of five representatives of labor, five representatives of management, and five public members, is responsibile for developing specific standards and criteria for pay adjustments consistent with the goals of the stabilization program. The Price Commission, consisting of seven public members, is established to develop standards and criteria for adjustments in prices and rents.

Wages, under the guidelines of the Pay Board, are to be limited to increases not in excess of 5.5 per cent. Meanwhile, price ceilings are set for coal producers (and manufacturing establishments) by the Price Commission at the level of the base prices in the base period, plus increases in costs so long as profit margins do not exceed the best two of the three fiscal years prior to August 15, 1971. The base period is the thirty days prior to August 15, 1971, and the base price is the highest price at which 10 per cent of transactions were made with a given class of customers during the base period. Wholesalers and retailers are to limit their markups to those in effect during the base period.

All enterprises are placed into one of three categories according to their volume of sales. Tier I firms, those with gross sales of $100 million or more, must have prior approval from the Price Commission of all price increases. Tier I firms must submit quarterly price, cost, and margin reports to the Price Commission. Tier II firms, those with gross sales between $50 million and $100 million, are not required to secure Price Commission approval for price increases, but must submit quarterly price, cost, and margin reports to the Commission, the same as Tier I firms. Tier III firms, those having less than $50 million gross sales, are not required to prenotify the Price Commission of price increases or submit quarterly reports. Nevertheless, the Tier III firms are governed by the same standards in pricing as the larger firms and are subject to monitoring and spot checks.

November 11, 1972—The Price Commission sets forth what it determines to be allowable costs for purposes of justifying price increases by producers of bituminous coal. Price increases that are subject to prenotification and approval will be governed by these standards, as will those of smaller firms not subject to prenotification. Long-term contracts containing flexible pricing formulas must also adhere to these guidelines. The allowable costs are:

1. Cost of wage increases falling within Price Commission standards, reduced to reflect an annual rate of productivity of 4.9 per cent:
2. Cost of increased employer contributions to:
   a) a pension, profit-sharing, or annuity and savings plan;
   b) a group insurance plan;
   c) a disability and health plan;
   d) old-age, survivors, and disability insurance; or
   e) an unemployment compensation fund;
3. Costs (including labor costs) to comply with state and federally mandated reclamation and mine health and safety requirements.

Phase III

January 11, 1973—Phase III regulations begin on this date and may be characterized as a general loosening of the strict Phase II program. The Price Commission and Pay Board are abolished under Phase III, and their responsibilities pass to the Cost of Living Council. Again, regulations affecting the coal industry follow the general provisions set by the Cost of Living Council. The general price standard is that prices may be increased over those authorized on January 10, 1973, so long as the profit margin does not increase. Prices may be increased, however, by a weighted average of 1.5 per cent over those permitted on January 10, 1973 to reflect increased costs without regard to profit margins. Prenotification of price changes is no longer required under Phase III. Controls are to be applied voluntarily on a self-administered basis. Quarterly reports on prices, costs, and margins are required from those firms with annual sales over $250 million.

June 13, 1973—The President announces a price freeze for not more than sixty days. Prices and margins are frozen at the highest level at which a substantial number of transactions occurred during the period June 1, 1973 through June 8, 1973. The purpose of the freeze is to halt the rapid increases in prices.

Phase IV

August 12, 1973—Phase IV general regulations become effective on this date at the end of the 60-day freeze period. General economic standards again include those controlling the coal industry. The base price for manufacturers (including coal producers) is set as the average price for the last fiscal quarter before January 11, 1973. Cost increases are allowed to be passed through only on a dollar-for-dollar basis. All firms with over $100 million in annual sales must once again give notice of any price increases. The markup of retailers and wholesalers cannot exceed that in effect during the base period. The base period for wholesalers and retailers is set at either the last fiscal year ending February 5, 1973 or the four consecutive fiscal quarters ending just prior to that date.

OPEC CHRONOLOGY

Early OPEC Activities

The Organization of Petroleum Exporting Countries (OPEC) was founded in 1960. Its original members were Iran, Iraq, Kuwait, Saudi Arabia, and Venezuela. In 1961 Qatar joined the Organization and was followed by Indonesia and Libya in 1962. Abu Dhabi and Algeria entered in 1967 and 1969 respectively. Nigeria became the eleventh member in 1971.

During the 1950's the concession agreements in producing countries called for the oil companies to pay host governments a tax on profits realized from the sale of exported oil. In paying a tax on profits, operating companies had to inform the governments of the receipts and outlays associated with their operations. In calculating receipts, the producing companies had to set a per barrel price on their oil. These prices were made public and were the basis on which the prevailing 50–50 profits tax was calculated. Thus, it can be seen that it was in the producing governments' best interest to keep these posted prices as high as possible. A higher posting would result in larger tax receipts for the landlord government.

From 1958 to 1960 the oil companies made several reductions in the posted price. The postings had been set too high, and market transactions were being made at prices discounted below prevailing postings. With the market price existing below the posted level, producing companies were paying income tax on income they were not realizing. This led to the reduction in posted prices.

OPEC's initial goal was to maintain the posted prices at the highest possible levels so as to receive larger income tax payments. OPEC succeeded in this goal throughout the 1960's. Price postings were effectively maintained throughout that period with no further reductions occurring.

This maintenance of posted prices above market prices changed the nature of the payments made by producing companies to host governments. The payments could no longer accurately be called an income tax, since they were based on an income that in reality was not being received by the oil companies. The companies were actually making a per barrel payment based on the fixed posted prices. Upon OPEC's successful maintenance of the posted price, it had thus become a tax reference price rather than a true market price.

Negotiations between OPEC members and producing companies in 1962 and 1963 led to larger government receipts. Royalties were no longer to be treated as income tax; they were now deductions from income. The following example will show how government revenues increased. A 50 per cent tax rate existed at this time. Let it be hypothetically assumed that gross profits are $50 and royalties $10. Under the old system, oil companies were liable for $25 in income tax; of this amount, $10 would be paid in royalties, while $15 would be paid in income tax. Under the new system, royalties are deducted from gross profits to set the new taxable base at $40 ($50 - $10 = $40). A 50 per cent tax rate sets the income tax at $20. Thus, payments to the government are now $30, $10 in royalties plus $20 in income tax. To ease the burden of expensing royalties, companies were allowed to make allowances off the posted price. The basic rate was set at 8.5 per cent in 1964, 7.5 per cent in 1965, and 6.5 per cent in 1966, after which the situation was to be reviewed.

Further negotiations in 1966 and 1967 called for the eventual phasing out of these allowances. By 1975, the gradual phase-out would result in tax payments being made on posted prices, less only the half-cent per barrel marketing allowance general in the Middle East.

Prior to 1966, three OPEC members (Venezuela, Libya, and Indonesia) still taxed the companies according to profits. This changed in 1967 when the countries agreed to set "tax reference prices" on the basis of which imputed profits for tax purposes would be calculated. In Libya, the shift was accomplished unilaterally by the government.

In 1968, OPEC issued its "Declaratory Statement of Petroleum Policy in Member Countries." This statement called for the renegotiation of existing contracts with the international oil firms. It was recommended that the posted prices be determined by the host governments and that prices should move in relation to the prices of manufactured goods traded internationally. Mention was also made of member countries' eventual participation in concessions.

OPEC Raises Posted Prices

Effective September 1, 1970, the first significant advance in crude oil posted prices occurred since the formation of OPEC. On this date, the posted prices of crude exported from Libya, Nigeria, and the Mediterranean ports handling Iraqi and Saudi Arabian crude oil rose by approximately 30 cents per barrel. The tax rate on these crudes was also raised from 50 per cent to a range of 54 to 58 per cent.

This round of increases marked the end of a long period of stable posted prices of foreign crude. In May 1970, the flow of oil through the Trans-Arabian Pipeline was stopped. This development, coupled with the closing of the Suez Canal by the 1967 Arab-Israeli War and with the growing demand for oil, led to a tanker shortage. With Mediterranean crude supplies curtailed, more Persian Gulf crude now had to travel the long route around Africa to get to market. Libya also had begun production cuts in an attempt to force operating companies to accept higher tax rates. This further aggravated the tanker shortage. Even more crude now had to move to market from the Persian Gulf around Africa. The tanker shortage was instrumental in raising prices and in increasing the bargaining power of the producing states.

Following the lead of the Mediterranean crude exporting countries, Persian Gulf producers soon made an upward revision. Iran and Kuwait, effective November 14, 1970, secured increases of 9 cents a barrel in postings for the 31° oil—to $1.72 and $1.68 respectively. At the same time, the tax rates in these two countries were raised to 55 per cent. Other Persian Gulf countries took similar actions.

In December 1970, the Venezuelan government took steps to increase its tax receipts. Legislation was introduced enabling the government to fix tax reference prices unilaterally. The tax rate was also raised from 52 to 60 per cent.

Teheran and Tripoli Agreements

On February 15, 1971, the Teheran Agreement was signed. Twenty-two international oil companies negotiated this agreement with six OPEC members (Iran, Iraq, Saudi Arabia, Kuwait, Abu Dhabi, and Qatar). These negotiations marked the first time that the oil companies bargained collectively with producing countries rather than individually. Major revisions in the pricing system resulted from the negotiations. General provisions provided for were: (1) a general increase of approximately 33 cents per barrel on posted prices; (2) upward adjustments on postings of heavier crude; (3) elimination of discounts remaining from the 1966–1967 agreements; (4) yearly incremental increases in posted prices through 1975; and (5) a minimum rate of taxation of 55 per cent.

The Tripoli Agreement, effective March 20, 1971, contained many of the provisions that were in the Teheran pact. Libya initiated the negotiations, and the agreement was also accepted by Nigeria and later by Saudi Arabia and Iraq for the portion of their exports piped to Mediterranean ports. The Tripoli increases were higher than the Teheran terms because Mediterranean exporters are closer to European markets. A low-sulfur premium, a Suez Canal closure premium, and a freight rate component were also added to the Tripoli pact.

The Geneva Agreements

The Teheran and Tripoli Agreements prevailed as written until early 1972. Then, the producing countries insisted on a change to to reflect the declining purchasing power of the U.S. dollar. This led to the so-called "first" Geneva Agreement. Under Geneva I, posted prices were to be raised 8.49 per cent effective January 20, 1972. Venezuela and Indonesia soon followed this lead and their posted prices also increased.

An amendment to the Geneva Agreement led to further increases in posted prices as of June 1, 1973. Following the U.S. dollar devaluation in early 1973, OPEC nations again sought increased compensation for their crude exports. The Geneva II terms established a parity index that was to compensate for any further fluctuations in the dollar. This new agreement resulted in an 11.9 per cent increase on June 1 postings over January 1, 1973 levels. The use of this index has resulted in several minor price adjustments since that time. Postings moved upwards on July 1 and August 1, downwards on October 1, increased on November 1, and decreased again on December 1, 1973.

General Agreement on Participation

A new influence on the pricing of crude oil arose in October of 1972. At that time, the General Agreement on Participation was signed by Saudi Arabia, Qatar, Abu Dhabi, and Kuwait and by the oil companies. However, the National Assembly of Kuwait later refused to ratify the agreement. This accord was the first major OPEC move towards a full working partnership in operating concessions. The oil firms' agreements with Abu Dhabi and Saudi Arabia were put into effect on January 1, 1973.

Under the terms of the agreement, the governments acquired an immediate 25 per cent interest in all crude oil and natural gas producing facilities within their borders. This includes exploration, development, production, gathering pipelines, and storage, delivery, and export facilities. Negotiations were also foreseen for the transportation and refinery facilities. The initial 25 per cent level was to be raised by fixed increments set on specific dates, rising to 51 per cent state control by January 1, 1982. The payments to be made to the oil companies on the share of the properties acquired by the host country were to be based on updated book values.

The agreement also attempted to minimize market disruptions by "participation" oil through the use of "bridging" and "phase-in" crude. During the first three years of the agreement, the governments were to sell back participation crude to the operating companies in order to help them "bridge" prior supply agreements. The companies were to pay an extra 33 to 57 cents per barrel over the tax-paid cost for this "bridging" crude. The oil companies also were to pay in excess of the tax-paid cost for the "phase-in" crude. "Phase-in" oil refers to the quantity of oil that the host governments required operating companies to buy back if the governments had difficulty in disposing of it themselves. The excess over the tax-paid costs resulting from the buy-back of "phase-in" crude would range during the first year from 21 to 52 cents per barrel. This buy-back participation crude would thus result in raising the costs on the whole of the operating companies' production.

Nigeria soon followed the lead of the Persian Gulf states and negotiated its own participation agreements. As of April 1, 1973, the government acquired a 35 per cent interest in the Shell-British Petroleum properties. The Nigerian government held the option to increase its share to 51 per cent by 1982. However, the rise to 51 per cent could be accomplished in any increments at any time before 1982.

Arab-Israeli War

The war between Isreal and the Arab nations of Egypt and Syria on October 6, 1973 led to the most dramatic events in the foreign crude market. The Arab members of OPEC agreed to use crude oil as a political weapon in the Arab-Israeli conflict, with all but one (Iraq) cutting back their oil production. Arab oil was embargoed to some countries, particularly the United States and the Netherlands. Pricing policies in effect under the 1971 Teheran and Tripoli Agreements were discarded. The producing countries declared themselves to be the unilateral determiners of crude oil postings. Effective October 16, 1973, the posted price of Arab crude oil rose by 70 per cent or more.

The cutbacks in production affected exports from Saudi Arabia, Kuwait, Abu Dhabi, Qatar, Libya, Algeria, Bahrain, Dubai, Egypt, and Syria. Supplies from non-Arab members of OPEC, such as Venezuela, Iran, Nigeria, and Indonesia were not affected. The original agreement on cutbacks, announced on October 17, 1973, called for a minimum 5 per cent production cut for October, with additional reductions of 5 per cent each succeeding month. Saudi Arabia, however, announced an immediate 10 per cent cutback. Other Arab nations took a variety of cutback actions.

Importing countries were graded according to their friendliness, neutrality, or hostility to the Arab political position. Hostile countries, the Netherlands and the United States, were the targets of an Arab oil embargo. Refining centers in the Caribbean, which supplied the U.S., were also cut off. Only Iraq chose not to follow the Arab embargo. However, Iraq did nationalize the concessions of the Dutch and American oil companies operating within the country.

Posted prices increased greatly. Arabian light crude jumped from an October 1, 1973 posting of $3.011 per barrel to $5.119 per barrel on October 16. Over the same time span, Iranian light crude rose from $2.995 to $5.091 per barrel. The freight advantage held by Libyan crude over Persian Gulf crude led to a greater increase in its posting. Libyan 40° crude rose from $4.604 to $8.925 per barrel from October 1 to October 16.

In January of 1974, Arab nations decided to reverse their production cutbacks by 10 per cent. However, the United States and the Netherlands remained on the embargo list.

Table 1. Estimated Cost of South Louisiana 36° Crude Oil Delivered to Selected Refinery Centers (Dollars Per Barrel)

	January 1, 1973	*October 16, 1973*	*January 1, 1974*
	Philadelphia		
Wellhead price[a]	3.67	4.27	5.27
Gathering charge	0.07	0.07	0.07
Pipeline charge	0.05	0.05	0.05
Tanker charge (charter)	0.50	0.66	0.70
Total delivered cost	4.29	5.05	6.09
	Houston		
Wellhead price[a]	3.67	4.27	5.27
Gathering charge	0.05	0.05	0.05
Pipeline charge	0.08	0.08	0.08
Total delivered cost	3.80	4.40	5.40
	Chicago		
Wellhead price[a]	3.67	4.27	5.27
Gathering charge	0.07	0.07	0.07
Pipeline charge	0.22	0.22	0.22
Total delivered cost	3.96	4.56	5.56

[a]The wellhead prices were postings for crude oil under price controls. The wellhead prices for exempt crude oil for January 1, 1974 were substantially higher with Platt's *Oilgram Price Service Crude Oil Supplement* stating that prices for exempt crude oil in the Texas Gulf Cost area as high as $10.40 per barrel had been announced.

Note: The estimated charter tanker charges are derived by utilizing spot tanker rates multiplied by 66. 7 per cent.

Source: Wellhead Prices—Platt's *Oilgram Price Service; pipeline and gathering charges—Digest of Pipeline Rates on Crude Petroleum Oil in Tariffs on File with the Interstate Commerce Commission,* compiled and published by Capital Systems Group Inc., Bethesda, Maryland.

Tanker Charges—Estimated from Platt's *Oilgram Price Service* and the Association of Broker and Agents Tanker Committee, *Association of Ship Brokers and Agents Freight Rate Schedules.*

Table 2. Estimated Cost of West Texas 36° Sweet Crude Oil Delivered to Selected Refinery Centers (Dollars Per Barrel)

	January 1, 1973	*October 16, 1973*	*January 1, 1974*
	Philadelphia		
Wellhead price[a]	3.48	4.23	5.23
Gathering charge	0.05	0.05	0.05
Pipeline charge	0.16	0.16	0.16
Tanker charge (charter)	0.50	0.66	0.70
Total delivered cost	4.19	5.10	6.14
	Houston		
Wellhead price[a]	3.48	4.23	5.23
Gathering charge	0.05	0.05	0.05
Pipeline charge	0.16	0.16	0.16
Total delivered cost	3.69	4.44	5.44
	Chicago		
Wellhead price[a]	3.48	4.23	5.23
Gathering charge	0.05	0.05	0.05
Pipeline charge	0.27	0.27	0.27
Total delivered cost	3.80	4.55	5.55

[a]The wellhead prices were postings for crude oil under price controls. The wellhead prices for exempt crude oil for January 1, 1974 were substantially higher with Platt's *Oilgram Price Service Crude Oil Supplement* stating that prices for exempt crude oil in the Texas Gulf Coast area as high as $10.40 per barrel had been announced.

Note: The estimated charter tanker charges are derived by utilizing spot tanker rates multiplied by 66.7 per cent.

Source: Wellhead Prices—Platt's *Oilgram Price Service; pipeline and gathering charges—Digest of Pipeline Rates on Crude Petroleum Oil in Tariffs on File with the Interstate Commerce Commission,* compiled and published by Capital Systems Group Inc., Bethesda, Maryland.

Tanker Charges—Estimated from Platt's *Oilgram Price Service* and the Association of Broker and Agents Tanker Committee, *Association of Ship Brokers and Agents Freight Rate Schedules.*

Major increases in posted prices were unilaterally imposed by OPEC governments in January 1974. Prices reached record levels and were to remain in effect throughout the first quarter of 1974. The posted price of Arabian light crude rose from a December 1 level of $5.04 per barrel to a January 1, 1974 level of $11.65 per barrel. Libyan 40° crude rose from $8.925 on October 19 to a posting of $15.768 per barrel on January 1. Venezuelan Oficina 35° crude posted price increased to a January 1 figure of $14.247 (excluding sulfur premium) from a November 1 level of $7.802 per barrel.

In setting posted prices for the entire first quarter of 1974, host governments have effectively scrapped the Geneva I and II Agreements of 1972 and 1973. The recent strength of the U.S. dollar would have resulted in approximately a 70 cents per barrel reduction in posted prices on January 1, 1974. In announcing that no price changes would occur through March 31, OPEC governments have eliminated any such adjustments that would have resulted under the terms of the Geneva Agreements.

Table 3. Estimated Cost of West Texas 36° Sour Crude Oil Delivered to Selected Refinery Centers (Dollars Per Barrel)

	January 1, 1973	*October 16, 1973*	*January 1, 1974*
	Philadelphia		
Wellhead price[a]	3.37	4.12	5.12
Gathering charge	0.05	0.05	0.05
Pipeline charges	0.16	0.16	0.16
Tanker charges (charter)	0.05	0.66	0.70
Total delivered cost	4.08	4.99	6.03
	Houston		
Wellhead price[a]	3.37	4.12	5.12
Gathering charge	0.05	0.05	0.05
Pipeline charges	0.16	0.16	0.16
Total delivered cost	3.58	4.33	5.33
	Chicago		
Wellhead price[a]	3.37	4.12	5.12
Gathering charge	0.05	0.05	0.05
Pipeline charges	0.27	0.27	0.27
Total delivered cost	3.69	4.44	5.44

[a] The wellhead prices were postings for crude oil under price controls. The wellhead prices for exempt crude oil for January 1, 1974 were substantially higher with Platt's *Oilgram Price Service Crude Oil Supplement* stating that prices for exempt crude oil in the Texas Gulf Coast area as high as $10.40 per barrel had been announced.

Note: The estimated charter tanker charges are derived by utilizing spot tanker rates multiplied by 66.7 per cent.

Source: Wellhead Prices—Platt's *Oilgram Price Service; pipeline and gathering charges—Digest of Pipeline Rates on Crude Petroleum Oil in Tariffs on File with the Interstate Commerce Commission*, compiled and published by Capital Systems Group, Inc., Bethesda, Maryland.

Tanker Charges—Estimated from Platt's *Oilgram Price Service* and the Association of Broker and Agents Tanker Committee, *Association of Ship Brokers and Agents Freight Rate Schedules*.

Table 4. Estimated Cost of Long Beach (Signal Hill) Huntington Beach 36° Crude Oil Delivered to Los Angeles Refineries (Dollars Per Barrel)

	January 1, 1973	*October 16, 1973*	*January 1, 1974*
Wellhead price[a]	3.72	4.32	5.32
Gathering charge and Pipeline charge	0.10[e]	0.10[e]	0.10[e]
Total delivered cost	3.82	4.42	5.42

[a] The wellhead prices were postings for crude oil under price controls. The wellhead prices for exempt crude oil for January 1, 1974 were substantially higher with Platt's *Oilgram Price Service Crude Oil Supplement* stating that prices for exempt crude oil in the Texas Gulf Coast area as high as $10.40 per barrel had been announced.

[e] Estimated.

Source: Wellhead Prices—Platt's *Oilgram Price Service; pipeline and gathering charges—Digest of Pipeline Rates on Crude Petroleum Oil in Tariffs on File with the Interstate Commerce Commission,* compiled and published by Capital Systems Group Inc., Bethesda, Maryland.

Table 5. Estimated Cost of Saudi Arabian Light Crude Oil 36° Delivered to Selected Refinery Centers (Dollars Per Barrel)

	January 1, 1973	*October 16, 1973*	*January 1, 1974*
	Philadelphia		
Cost of crude oil-Ras Tanura[a]	1.95	3.68	8.43
Tanker charges (charter)	1.14	1.82	1.93
Duty or license fee	0.105	—[b]	—[b]
Total delivered cost	3.195	5.50	10.36
	Houston		
Cost of crude oil-Ras Tanura[a]	1.95	3.68	8.43
Tanker charges (charter)	1.19	1.92	2.02
Duty or license fee	0.105	—[b]	—[b]
Total delivered cost	3.245	5.60	10.45
	Chicago		
Cost of crude oil-Ras Tanura[a]	1.95	3.68	8.43
Tanker charges (charter)	1.18	1.89	2.00
Duty or license fee	0.105	—[b]	—[b]
Pipeline charges	0.22	0.22	0.22
Total delivered cost	3.455	5.79	10.65
	Los Angeles		
Cost of crude oil-Ras Tanura[a]	1.95	3.68	8.43
Tanker charges (charter)	1.10	1.77	1.87
Duty or license fee	0.105	—[b]	—[b]
Total delivered cost	3.155	5.45	10.30

[a] These cost estimates represent the tax paid cost (government take and cost of production) of the producing company's equity crude adjusted from the reported 34° gravity to 36° gravity plus the company margin, as indicated by the *Petroleum Intelligence Weekly,* and may be considered as the minimum price of the crude F.O.B. embarkation point. The cost of "buy-back" (crude oil) the company gets from the government together with the equity crude it produces is not known but undoubtedly will be in excess of the tax paid cost on equity oil.

[b] Assumes fee free crude oil imported within allocations (fee free crude oil comprises most of the imports).

Source: Cost of crude oil, Ras Tanura—*Petroleum Intelligence Weekly,* Vol. XIII, No. 5, February 4, 1974, pages 2 and 4; *Petroleum Intelligence Weekly Supplement,* October 8, 1973, page 4.

Tanker charges—Association of Ship Brokers and Agents, Inc., *Worldwide Tanker Nominal Freight Scale;* Afra Tanker Rates for previous month published by the *Petroleum Economist* (formerly the *Petroleum Press Service*).

Duty or license fee—U.S. Department of Interior, Office of Oil and Gas, *World Wide Crude Oil Prices, Summer 1972,* and *World Wide Crude Oil Prices, Summer 1973.*

Part II

Basic Study

Introduction and Summary

This Part of the report was prepared by Foster Associates, Inc. (Washington, D.C.) for the Energy Policy Project of the Ford Foundation. It was begun in March 1972 and completed in August 1972.

Purpose and Parameters of Study:

The purpose of this study is to determine trends since 1960 in prices of important primary and secondary sources of energy in the United States and to provide basic price data useful in the analysis of various other aspects of the nation's energy sources.

Price data, both retail and wholesale, have been compiled in this study for the following major energy sources: natural gas, distillate fuel oil, residual fuel oil, motor gasoline, coal, and electricity. For natural gas, electricity, and motor gasoline the price data cover twenty-four standard metropolitan statistical areas (SMSAs) distributed over twenty-two states and the District of Columbia, with all Census regions represented. Prices for fuel oil and coal, however, are limited to those of the twenty-four SMSAs where these fuels are a significant source of energy. In addition to providing energy prices for a broad cross-section of the Nation, the study also covers price data for energy in most of its major uses—residential, commercial, industrial, power plants, and automobiles.

The period covered by the price data is from 1960 to 1972, with early 1973 prices indicated in some categories. Generally, the prices are for two dates in each year, around January 15 and July 15, but for some price series it was practical to give only annual averages.

This study of energy prices in the United States is largely devoted to the compilation of retail and wholesale prices of major primary and secondary sources of energy. The bulk of the report consists of tables on energy prices, and a substantial portion of the text is devoted to a description of those basic elements essential to the understanding and use of the tables, including: (1) the industry structure of major energy sources, (2) the definitions and criteria used in developing the price information, (3) the methodology, and (4) sources of price data. A relatively brief description is given of energy price trends and patterns in a broad cross-section of SMSAs, and the remainder of the report consists of special analyses of different aspects of energy prices.

INDUSTRY STRUCTURE OF MAJOR ENERGY SOURCES

The gas and electric industries consist largely of investor-owned companies whose prices are regulated by government agencies. In the gas industry there are three sectors: (1) the exploration and production sector, which comprises major oil companies and independent operators; (2) the transmission sector, which transports the gas to the market and is composed largely of independently owned pipelines; and (3) the distribution sector, made up primarily of local utilities, which retail the gas in limited markets. The prices of gas flowing in interstate commerce are regulated by the Federal Power Commission, which controls both the field prices of gas and the pipeline transportation costs. In intrastate commerce, however, the field price of gas is not regulated and currently tends to be higher than the regulated wellhead price of gas in interstate commerce. Unlike gas, most electricity is generated, transported, and distributed by the same company to a market within one state or part of a state, and its price is regulated by state utility commissions.

Prices of oil products are determined by market forces of supply and demand, although the government has had an important influence on the prices—directly in times of general price controls and indirectly through measures that affect supply and demand. The oil industry is more complex than the gas and electric industries. It comprises four sectors: exploration and production, refining, transportation, and marketing. Some of the major oil companies are largely integrated, operating in all four sectors; others are integrated in varying degrees. Nevertheless, there are many independently owned operators or companies operating in only one or sometimes two of the four sectors.

While the oil prices are set by market forces, government agencies regulate or influence each of the four sectors of the industry. Moreover, insofar as government regulatory bodies affect prices of forms of energy other than oil, they also have some influence on oil prices.

Coal prices also are largely determined by market forces. Because of the nation's wide distribution of coal resources and the relatively high cost of transporting it, coal markets tend to be regional.

Most steam coal now is sold by the producer directly to the ultimate consumer—typically a power plant or an industrial plant. Steam coal is used primarily for its heat content and its price is influenced by competition from other fuels.

SELECTION OF ENERGY SOURCES, USERS, AND GEOGRAPHIC LOCATIONS FOR PRICING

The study was designed to cover prices for the nation's major sources of energy since 1960—gas, electricity, #2 fuel oil, #6 fuel oil, motor gasoline, and coal; and for the major types of users—residential, commercial, industrial, and power plants. Meanwhile, the twenty-four SMSAs for which prices were compiled for the different sources of energy for the different types of users were determined on the basis of population and geographic distribution to enable the analysis of any fuel price patterns that might result from regional and area economic and climatic factors or from accessibility to different modes of transportation.

In some of the twenty-four SMSAs more than one gas or one electric utility served the area. In these instances the utility supplying the bulk of the customers their gas or electricity service was chosen for the study.

DEFINITIONS AND CONCEPTS

Both the retail and wholesale prices for the different forms of energy were included in the study. Retail prices were defined as those prevailing in the sales to the ultimate customer or user of the energy. Thus, sales to a residential customer and to a power plant were considered retail sales. Wholesale prices were defined as the prices paid by the retailer to his supplier.

In residential heating and air conditioning, the prices were based on quantities of fuels needed if normal temperatures prevailed in each of the SMSAs. For other retail prices—residential baseloads, small commercial, large commercial, large industrial (but not for power plants)—the same quantity for each type of user was used for a particular type of energy in all the SMSAs. Except for electricity, the quantities used for the different kinds of fuel were equivalent to meet the given requirements of the customer. Moreover, for all users except power plants, the quantities of the different types of energy were held unchanged throughout the study period 1960—1972. Thus, the price series on the various energy sources for the different types of use reflect only price changes, and not changes due to volume changes.[a]

The retail prices are delivered prices and include excise and sales or use taxes. It should be pointed out, however, that the fuel oil prices for large commercial and large industrial users are based on posted prices, which generally deviate from contract prices under which moves the major portion of the fuel oil to these customers. The wholesale prices vary due to the different structures of the energy industries, with gas being the price delivered to the city gate; gasoline the price of the product delivered to the dealer's station; fuel oil the terminal price; electricity the estimated cost of generation; and coal the price at the source.

All prices in the study are stated in current dollars, with no adjustment being made for changes in the value of the dollar. The prices are usually for the volumetric units appropriate to the particular form of energy. The prices for residential and small commercial users, however, are in terms of the average monthly bills. Moreover, in a few tables on industrial and power plant use, where direct comparisons of alternate fuel costs are made, prices are shown in cents per million Btu for all fuels.

DATA SOURCES

Most of the energy prices were compiled from published sources, but some data were obtained from energy suppliers. The principal sources were as follows:

Gas—the American Gas Association's *Rate Service,* Form 2, *Annual Reports of Interstate Pipelines to the Federal Power Commission.*

Electricity—the Federal Power Commission's *National Electric Rate Book,* Form 1 and 1-M, *Annual Reports of Electric Utilities to the Federal Power Commission.*

Oil Products—*Platt's Oilgram Price Service,* the Bureau of Labor Statistics' *Retail Fuel Prices and Indexes of Fuels and Electricity, Fuel Oil and Oil Heat, The Oil and Gas Journal,* and data supplied by various oil companies.

Coal—data provided by coal producers, customers, and the Bureau of Mine's *Minerals Year Book.*

[a]It was not practical to follow the same procedure on power plants, the fuel prices of which represent annual averages paid by the different utilities for each of the various fuels used each year.

TRENDS AND PATTERNS IN ENERGY PRICES

Energy prices, both retail and wholesale, tended to be relatively stable during most of the 1960's, but in recent years they have shown marked and widespread increases. Prior to 1969, electricity prices declined while gas prices mostly registered only modest changes, with increases in some SMSAs balanced by decreases in others. Prices of oil products and coal, however, began to rise around the middle of the 1960's and have continued to rise up to the present.

The Northeast section of the nation has continued to show the highest energy prices. Conversely, the West South Central region, the nation's major source of gas and oil, has continued to experience the lowest energy prices.

While all types of users have had an increase in their energy costs in recent years, the relative increases for the large commercial and large industrial users has been somewhat greater for gas and electricity than for the small user. This pattern did not appear to prevail for distillate fuel oil, at least from 1966 to 1972. Since residual fuel oil and coal are mainly consumed by large users, the comparison between large and small users is not relevant, but prices of these two fuels have shown greater increases in recent years than have other energy sources.

BREAKDOWN OF RETAIL ENERGY PRICES BY COST COMPONENT

A breakdown of the retail energy prices into their various cost components reveals that the production component of retail prices in 1971 was higher than in 1960 in most SMSAs for all the major energy sources. Excise and sales taxes also generally rose during this period. On the other hand, transportation and distribution costs of gas and electricity showed a mixed pattern for the different SMSAs and for the different types of users. Transportation and distribution costs of fuel oil, gasoline, and coal generally were higher in 1971 than in 1960.

Reflecting variations in the relative changes of the different cost components, the share of the total retail prices accounted for by the various cost components changed somewhat for the various forms of energy among the different types of users. For gas and electricity, production costs represented a larger portion of the retail price to most types of users in 1971 than in 1960 in a preponderance of the SMSAs. Coal showed the same pattern in the price to large industrial users. Meanwhile, the production component in the price of oil products to the various types of users in 1971 represented a somewhat smaller portion of the total retail price than in 1960. Taxes increased in relative importance in the price of all types of energy to all types of users, except for retail gasoline.

Transportation and distribution costs generally made up a larger portion of the retail price paid by the small volume consumer such as residential, small commercial, and retail gasoline consumers than for large users such as large commercial and industrial establishments. Among the different forms of energy, the production component generally represented a larger portion of the total retail price of oil products than of other energy forms. Meanwhile, production costs were an appreciably smaller part of gas retail prices than of other types of energy. Oil products tended to have relatively low transportation costs, while gas tended to have relatively high transportation and distribution costs.

COMPARISON OF RURAL AND URBAN SPACE HEATING AND GASOLINE PRICES

Analysis indicates that prices for space heating in towns near metropolitan centers are not consistently higher or lower than those in metropolitan centers. For #2 fuel oil, prices in towns were generally the same as in metropolitan centers in 1972. Although electric and gas rates for space heating were higher in metropolitan centers than in nearby towns in most cases in 1972, the reverse situation prevailed in a substantial number of other cases. Moreover, the difference between metropolitan center and nearby town gas prices was relatively small in most cases, and the difference between electricity rates was narrowing between 1960 and 1972. For retail gasoline, the prices in towns appear to have been generally the same as in the nearby metropolitan centers, if allowance is made for distortions resulting from periods of depressed pricing.

RELATIONSHIP OF PRICES TO VOLUMES

Prices of all the major forms of energy tend to decrease as the quantity purchased increases. This phenomenon was demonstrated in an analysis of the price-volume relationship existing for the various forms of energy sold in the Chicago markets in December 1972. The same tendency exists in other markets. The price-volume relationship, however, is not strictly linear. Small quantities command a more than proportionately higher price than large quantities. On the other hand, the price-volume relationship tends to flatten out for very large volume purchases. Differences in the cost per Btu among the various fuels is substantially larger for small quantity purchases than for large.

VOLUME DISCOUNTS AND PROMOTIONAL RATES

Electric and gas utilities offer a variety of promotional rates. These rates provide lower unit costs to the customer either through increases in his consumption at any particular period of time or by increases in his consumption during periods or seasons of the year when his consumption normally would be low.

Utility rates are usually structured so that a constant price is charged for all energy consumed in the first 'block' and a successively lower price is charged for all energy consumed in additional blocks. Thus, promotional rates that encourage increased consumption at any period of time are in effect volume discounts. Promotional rates that tend to increase a customer's consumption during periods when it normally would be low are designed to give the utility a higher average load factor. A higher average load factor is important to a utility because of its heavy fixed costs.

Some of the types of promotional rates available are: for gas—a summer-winter rate with lower summer charges and an all-gas home rate where the customer must have both space heating and water heating; and for electricity—a summer-winter rate with the winter charges lower and lower rates for water heating and/or space heating.

INTERRUPTIBLE GAS RATES FOR INDUSTRIAL MARKETS

Interruptible industrial gas rates pertain to sales in which the provision is attached that the service can be curtailed to accommodate the needs of firm customers, usually in the winter. These rates generally are significantly lower than the rates on firm sales. The user of interruptible gas must have alternate fuel burning facilities to meet his needs when his gas supply is curtailed. The gas distributor has utilized interruptible rates as a means of promoting the use of gas in off-peak periods or seasons as a means of maximizing the utilization of his facilities and accordingly lowering his total unit costs, a significant portion of which are fixed costs.

Interruptible gas rates have tended to be very competitive with the prices of alternate fuels. Moreover, the spread between the rates for interruptible and those for firm gas have tended to be wider in SMSAs where the competition from other fuels has been very keen. This spread has tended to narrow during the past twelve years.

SPOT VERSUS CONTRACT PRICES

The major portion of the fuel oil and coal marketed is sold on a contract basis covering a period of one or more years. Based on the volume sold, the contract price—which in large volume sales is generally a negotiated price—is more representative of the price at which the fuel oils and coal are moving than is the spot price. The spot price, however, is a more sensitive indicator of market conditions than the contract price and registers wider swings. During the relatively tight fuel supply situation during the 1973-1974 winter, the spot prices for fuel oils and coal were substantially higher than the contract prices.

INCREMENTAL GAS PRICING

The incremental pricing concept is that retail prices be based on the specific cost of each incremental source of supply. This concept contrasts with the prevailing practice of rolling in the cost of incremental supplies with the cost of other supplies. Gas prices are based on the average costs of all supplies.

The Federal Power Commission brought the incremental pricing concept to the fore in a decision, issued on June 28, 1972 (Opinion 622), authorizing three interstate pipelines to import LNG from El Paso Algerian Corporation, but imposing several conditions—one of which was the use of incremental pricing for this imported LNG not only by the pipeline but also by the distributors to whom the pipelines sold the incremental gas. Costs of the LNG would be substantially higher than the pipelines' existing supplies. In a subsequent action (Opinion 622A, issued October 5, 1972), the FPC eliminated the condition that distributors would have to price this gas on an incremental rate basis.

Chapter One

Industry Structure of Major Energy Sources

A. GAS AND ELECTRICITY

In the United States, the gas and electric industries predominantly consist of investor-owned companies.[a] The prices of gas and electricity sold by these companies are regulated by government agencies—the Federal Power Commission, state public utility commissions, and in a few instances, by municipal government utility commissions.

The gas industry breaks down into three sectors: (1) the exploration and production sector; (2) the transmission sector, and (3) the distribution (retailing) sector. Most of the gas produced in the United States is located in the Southwest (primarily in Texas, Louisiana, Oklahoma, and New Mexico). This gas production is transported by pipeline companies from the producing fields to market areas throughout the United States. The field price of gas that is transported in intrastate commerce is not regulated. It is determined by market factors, as evidenced by gas contracts between producers and transporters.[b]

Field prices of gas transported in interstate commerce have been regulated by the Federal Power Commission since 1954. The Federal Power Commission also regulates the cost of transporting gas shipped

[a]Government-owned electric and gas distributors are primarily located in the Southeast and Northwest portions of the country. The role of government-owned facilities in gas production and gas transmission is negligible.

[b]Gas-producing companies are also oil-producing companies, for gas and oil are typically found in the same areas and often in the same fields. These producing companies typically transport oil to refineries and thence to the market. However, most gas production is typically shipped to the market by independently-owned pipelines that, in turn, typically sell to independently-owned gas distributors.

interstate from the field to the "city gate" of gas distributors. The price of distributing this gas to the consumer is typically an intrastate commerce function and is regulated by state or municipal commissions.

The electric utility industry is different from the gas industry, and their price processes differ as well. Most electric utilities generate, transport, and distribute electricity within one state and therefore are subject to regulation by the state public utility commission.[a] This price regulation is typically focused on the price of electricity as sold by the utility to the consumer, which is determined by the sum of the cost of generating, transmitting, and distributing the electricity.

B. PETROLEUM

Unlike the price of gas and electricity, the price of crude oil and petroleum products has not been directly regulated by government agencies, except in times of general price controls. Prices of petroleum have been determined by market forces of supply and demand. The overall supply of petroleum, however, has been significantly affected by a variety of government policies—both federal and state. The demand for petroleum has been influenced by governmental regulation of the price of other energy sources, particularly of natural gas.

The petroleum industry may be divided into four major segments: (1) exploration and production; (2) refining; (3) transportation; and (4) marketing. The structure of each of these segments varies considerably.

The petroleum industry, on the whole, is a capital intensive industry. The large investments involved have tended to promote integration in the ownership of all segments as a means for interrelating and utilizing on a scale basis the investment in any one segment. Hence, large integrated oil companies have emerged, some of which are international in scope. However, the petroleum industry also includes thousands of firms in lesser degrees of integration.

In the exploration and production segment of the petroleum industry, most of the exploratory wells historically have been drilled by independent producers. A substantial portion of the oil produced and of the gathering systems that purchase the oil produced by the various wells in an oil field are owned by the integrated oil companies. The company gathering the crude oil posts the price it will pay for the crude; but if this price should become out of line with postings for other comparable crude, or if crude becomes tight, the posting must be changed to forestall the loss of connections.

[a]The Federal Power Commission regulation of electric prices is primarily limited to the relatively small amount of electricity sold wholesale in interstate commerce.

Of the nation's 256 operating refineries, about two-fifths are owned by major integrated oil companies and the remainder owned by others, including independent refiners, some of which are partially integrated companies. Since the refineries of the integrated majors tend to be larger than those of the independent refiners, the majors own a larger share of the national refining capacity. Companies are encouraged to operate their refineries as near to capacity as is practical because of the large investment and high fixed costs in refinery operations. While most of the refinery output of major companies is for their own marketing outlets, major companies frequently sell some of their refinery output to independent marketers. Independent refiners also sell some of their output to independent marketers, in addition to marketing some on their own and selling some to integrated major oil companies.

The transportation segment of the petroleum industry includes pipelines, water (tankers and barges), trucks, and rail. Integrated oil companies frequently own—individually or in joint ventures—crude-oil as well as product pipelines, but some pipelines are owned by independent operators or non-oil companies such as railroads. Pipeline tariffs are subject to government regulation. Major oil companies own tankers and barges, but independent tanker operators and barge lines also carry a significant amount of the oil moved, including major oil company shipments. Truck transportation of the petroleum industry is distributed over a large number of different operators, some of them directly engaged in marketing petroleum products and some of them common carriers. The common carrier truck as well as rail tariffs for moving petroleum are subject to government regulation.[a]

The marketing segment of the petroleum industry has a widely varying structure. Among the variety of participants in the marketing function are branded and unbranded jobbers and dealers, distributors, commission agents, and brokers Price wars historically have been frequent and widespread in gasoline retailing, the largest single segment of petroleum marketing. While oil marketers generally post selling prices for their various products, the prices of the bulk of the products moving to larger commercial, industrial, and government consumers are on negotiated contracts or on bids, which generally vary from the postings.

C. STEAM COAL

The coal industry is structured primarily around the producing function, although some coal companies also have marketing and transportation functions. For the smaller companies that do not market their own coal,

[a]Coking coal, which is used in the steel-making process and can be considered an industrial feedstock, is not a part of this study.

a broker is used to represent a number of producers seeking a market. With the demise of the residential market, very little coal is sold through the so-called "retail yards" or distributors. Most coal is now sold directly to the consumer.

Steam coal is used primarily for its heat content. The reserves of steam coal are widely distributed throughout the United States; hence, the markets for steam coal have tended to be regional. This is due to the fact that coal is costly to transport because of its bulk. A consumer would therefore purchase coal from the nearest source to minimize the cost.

The power plant and industrial sectors are the primary markets for steam coal. The power plant sector has become the largest user of coal, accounting for approximately 66 per cent of total coal used in 1971.

The industrial sector, which has been decreasing in importance, accounted for 14 per cent of total coal use in 1971.[a] Both of these market sectors purchase coal on a contract or "spot" basis. The term of the contract can be anywhere from a year, as in the case of an industrial user, to twenty years, as in the case of the power plant purchasing coal on a long-term contract basis. In addition to buying on a contract basis, users will also buy on a "spot" basis to supply a certain portion of their fuel requirements. As the name implies, buying on a spot basis involves purchasing what is available at any point in time at the prevailing price.

In its simplest form, the delivered price of coal can be broken down into the price at the mine or F.O.B. mine price, plus the transportation cost to the ultimate users. The F.O.B. mine price, in addition to the cost of mining the coal, will reflect the quality and size of coal required by the user. If delivered on a contract basis, the F.O.B. mine price frequently will be tied to an escalation provision that adjusts the price of coal for changes in the cost of mining it. Transportation costs are dependent upon the tariffs of the common carriers (who fall under the regulation of the Interstate Commerce Commission) in question. Barge movements of coal, however, are not regulated; their transportation costs are negotiated contract rates.

[a]The remaining 20 per cent was distributed between household use (3 per cent) and coking coal (17 per cent).

Chapter Two

Criteria, Definitions, and Concepts

A. SELECTION OF THE CITIES TO BE STUDIED

The twenty-four Standard Metropolitan Statistical Areas (SMSAs) for which data have been tabulated have been chosen to provide a reasonable cross-cut of the U.S. and to demonstrate different price structures resulting from such factors as (1) population density, (2) regional location, (3) climate, (4) economic influences, (5) different patterns of energy consumption, and (6) accessibility to navigable water. All census regions are represented.

B. MARKETS PRICED FOR PARTICULAR ENERGY SOURCES

Gas and electricity have been priced for all areas and market segments at the retail level. Retail sales are considered to be those to ultimate customers. They include applicable levels for household users as well as commercial and industrial customers. Since most of the steam-electric generating plants of electric utilities consume only fossil fuels, the cost of gas, as well as of coal and oil, has been presented for this sector in the tables.

There is complete tabular coverage of gas and electricity at the wholesale level, on either an actual or an imputed basis. Wholesale transactions cover those between a supplier and the final distributor. It is possible for these roles to overlap, i.e., for an electric power company to generate electricity and then sell it.

Coal and oil have been priced for those SMSAs and market sectors where their use has been determined to be significant. On the basis of data published in the 1970 Census of Housing, the percentage of occupied dwelling units using each of these fuels for house heating purposes was calculated for the SMSAs under study. Since in all twenty-four SMSAs the percentage was less than 10 per cent for coal, it has been excluded from the residential sector. On the same basis, #2 fuel oil has been priced in only ten of the twenty-four SMSAs for this type of customer.

For commercial establishments, the use of coal and fuel oil is assumed to follow generally the same pattern as in the residential sector. Hence, coal would not be significant in any of the twenty-four SMSAs in the study, and accordingly was not priced. #2 fuel oil was priced for both small and large commercial use in the same ten SMSAs as for residential use. Meanwhile, #6 fuel oil, which for the most part is used only for large installations located in the general area of oil refineries or near low cost water transportation, was priced only for the large commercial establishments in the nine SMSAs reasonably accessible to a source of supply.

Coal and fuel oil used by large industrial plants were priced for SMSAs where these fuels supply 10 per cent or more of the energy used by manufacturing industry. Reliable, definitive, current data on manufacturing industry fuel consumption by SMSAs are not available, and the selection of the SMSAs for pricing was based on fragmentary information from a variety of sources, including *Census of Manufacturers* and *Annual Survey of Manufacturers* of the U.S. Bureau of Census, the Bureau of Mines, the American Gas Association, the National Coal Association, and oil and coal companies. As a result of this investigation, it was determined to price coal for industrial plants in fifteen of the twenty-four SMSAs of the study, and to price #2 fuel oil for ten and #6 fuel oil for eleven of the SMSAs. Some SMSAs may have been included where these fuels may not provide as much as 10 per cent of the energy requirements of manufacturing plants, but it is unlikely that any SMSAs have been omitted where these fuels exceed 10 per cent of the total manufacturing energy consumption.

Wholesale prices of coal and fuel oils have been developed for the same SMSAs for which the retail prices of these fuels have been compiled. The wholesale price and the retail price of motor gasoline have been included for all the twenty-four SMSAs.

C. CHOICE OF UTILITIES

Where an SMSA was served by two or more utilities, the distributor supplying the bulk of the customers was chosen for study. For analytical purposes, both an investor-owned and a publicly-owned electric company were selected for three cities—Omaha, Nebraska; Los Angeles, California; Seattle, Washington.

The following SMSAs and gas and electric utilities were selected:

Census Region SMSA	*Gas Utility*	*Electric Utility(s)*
New England		
Boston, Mass.	Boston Gas Co.	Boston Edison Co.
Middle Atlantic		
Albany, N.Y.	Niagara Mohawk Power Corp.	Niagara Mohawk Power Corp.
New York, N.Y.	Brooklyn Union Gas Co.	Consolidated Edison of N.Y., Inc.
Pittsburgh, Pa.	Equitable Gas Co.	Duquesne Light Co.
South Atlantic		
Atlanta, Ga.	Atlanta Gas Light Co.	Georgia Power Co.
Charlotte, N.C.	Piedmont Natural Gas Co.	Duke Power Co.
Washington, D.C.	Washington Gas Light Co.	Potomac Electric Power Co.
Miami, Fla.[a]	Florida Gas Light	Florida Power & Light
East South Central		
Nashville, Tenn.	Nashville Gas Co.	Nashville Electric Service
East North Central		
Chicago, Ill.	The Peoples Gas, Light & Coke Co.	Commonwealth Edison Co.
Detroit, Mich.	Michigan Consolidated Gas Co.	Detroit Edison Co.
Columbus, Ohio	Columbia Gas of Ohio Inc.	Columbus & Southern Ohio Electric Co.

[a]The Houston Corporation was the distributor in 1960 and 1961.

West South Central		
Houston, Tex.	Houston Natural Gas Co.	Houston Lighting & Power Co.
New Orleans, La.	New Orleans Public Service, Inc.	New Orleans Public Service, Inc.
Tulsa, Okla.	Oklahoma Natural Gas Co.	Public Service Co. of Okla.
West North Central		
Minneapolis-St. Paul, Minn.	Minneapolis Gas Co.	Northern States Power Co.
Omaha, Neb.	Metropolitan Utilities District	Omaha Public Power District Iowa Power & Light Co.
St. Louis, Mo.	Laclede Gas Co.	Union Electric Co.
Mountain		
Denver, Colo.	Public Service Co. of Colo.	Public Service Co. of Colorado
Phoenix, Ariz.	Arizona Public Service Co.	Arizona Public Service Co.
Salt Lake City, Utah	Mountain Fuel Supply Co.	Utah Power & Light Co.
Pacific		
Los Angeles,	Southern California Gas Co.	Los Angeles Dept. of Water & Power Southern California Edison Co.
San Francisco, Calif.	Pacific Gas & Electric Co.	Pacific Gas & Electric Co.
Seattle, Wash.	Washington Natural Gas Co.	Puget Sound Power & Light Co. Seattle Dept. of Lighting

D. DEFINITION OF CONCEPTS AND DETERMINATION OF QUANTITIES TO BE MEASURED—RETAIL PRICES

The figures appearing on the residential tables represent the monthly bills for the identical services provided by alternate types of fuels. The electric non-substitutable baseload, which covers expenditures for lighting, small appliances, and such large appliances as the refrigerator, and which is described in greater detail below, is the one exception, since neither gas nor fuel oil compete as a source for these uses. However, for the substitutable baseload, which includes the water heater, clothes dryer, and range as well as space heating and cooling, alternatives to electric power are common and are included in the study.

The services to be measured are based on an equivalent Btu volume in order to make possible meaningful comparisons between the different types of energy. Quantities have been held constant throughout the period 1960 to January 1973 so that the resultant series reflect only price changes. For gas, electricity, and #2 fuel oil the outlays presented represent use in a normal winter month, with gas and electricity priced according to rates prevailing on January 1 and July 1 in the given year and oil based on the average of prices in effect on January 15 and July 15 of that year and on January 15 of the succeeding year.

Non-Substitutable Electric Baseload

This block of energy consumption covers the use of small appliances and certain large appliances in addition to lighting. The Housing Census was checked to determine the average number of rooms and baths in occupied dwelling units in SMSAs for 1970. An estimate of the wattage and hours of use for each room was derived. To this lighting component was added a figure representing the monthly power use of small appliances (i.e. coffeemaker, radio, iron, etc.) and of large electric appliances having no significant gas counterpart (the refrigerator, clothes washer, dishwasher, and television). Kilowatt consumption of each of these products was taken from Edison Electric Institute data and adjusted by a saturation index taken from *Merchandising Week* based on the percentage of U.S. homes with one or more of these products in order to determine a typical mix of appliances for the average home. The resulting calculations suggested 400 kilowatt hours (kwh) per month as a reasonable level of power use for this component.

Substitutable Baseload

Four thousand cubic feet (4 Mcf) of gas and 800 Kwh of electricity have been established as appropriate volumes for this portion of the bill. This covers power consumed in the operation of a range, clothes dryer, and water heater. Disparate estimates of the electric power consumption of these appliances for a given period of time from Edison Electric Institute data and selected operating electric utilities stemmed from different assumptions as to their hours of use. Figures for the gas consumption of these products from the American Gas Association, the Independent Natural Gas Association of America and selected operating gas utilities varied within a much more narrow range and suggested a volume of 4 Mcf per month, with approximately 65 per cent accounted for by the water heater.

In order to hold the hours of use constant for both electric and gas users, it was decided to convert this gas volume, for which there seemed to be general agreement to a Kwh equivalent. Comparative efficiency factors for gas and electric water heaters and ranges were determined from the *Gas Engineers Handbook,* 1965. No data were available for dryers; consequently, the same efficiency ratio was assumed as for the other two appliances. The Btu content of the volume of electricity required to perform a unit of service was approximately 67 per cent of the Btu content of gas required to perform an equivalent service. This percentage was applied to the monthly gas consumption in Btu's for each of the three appliances and converted to Kwh's. The comparable electric baseload was estimated to be about 800 Kwh. The average daily use implicit in these assumptions are .9 hour for the dryer, .5 hour for range, and 3.8 hours for the water heater.

Space Heating

The volume of gas consumption for space heating was obtained from annual issues of the INGAA[a] publication entitled *Comparison of Seasonal Househeating Costs for Gas, Fuel Oil, Coal and Electricity,* 1960 to 1972. For each of the five SMSAs in this study for which data were not available, an estimate was made by calculating the ratio of its normal heating degree days[b] to the normal degree days in the geographically nearest SMSA for which data were available, and by applying the ratio to the average volume consumed during the heating season in the nearest SMSA.

The justification for this procedure is indicated in the accompanying scatter diagram (Figure 2–1) showing the relationship between degree days and the volume of fuel consumed for heating. The scatter around this average line of relationship can be largely attributable to such construction factors as the quality of insulation, and to weather conditions such as wind, the amount of sunlight, etc.

Having established the volume of gas for space heating during a normal winter month for each SMSA, its kilowatt equivalent was determined, using a thermal efficiency factor of 75 per cent. This represents the ratio of the usable energy output of gas to that of electricity for an identical Btu input and is taken from an average of factors used by the distributors reporting in the INGAA tabulation.

The quantity of #2 fuel oil for space heating during a normal month for each SMSA was derived by converting the volumes used for gas to equivalent volumes for fuel oil, which then were raised 5 per cent to adjust for the difference in thermal efficiency between the two fuels. These thermal efficiency factors were based on studies of the Independent Natural Gas Association of America.

[a]Independent Natural Gas Association of America.

[b]Heating degree days measure the extent to which the mean daily temperature falls below 65 degrees. One degree day is counted for each degree of deficiency for every day on which such a deficiency occurs.

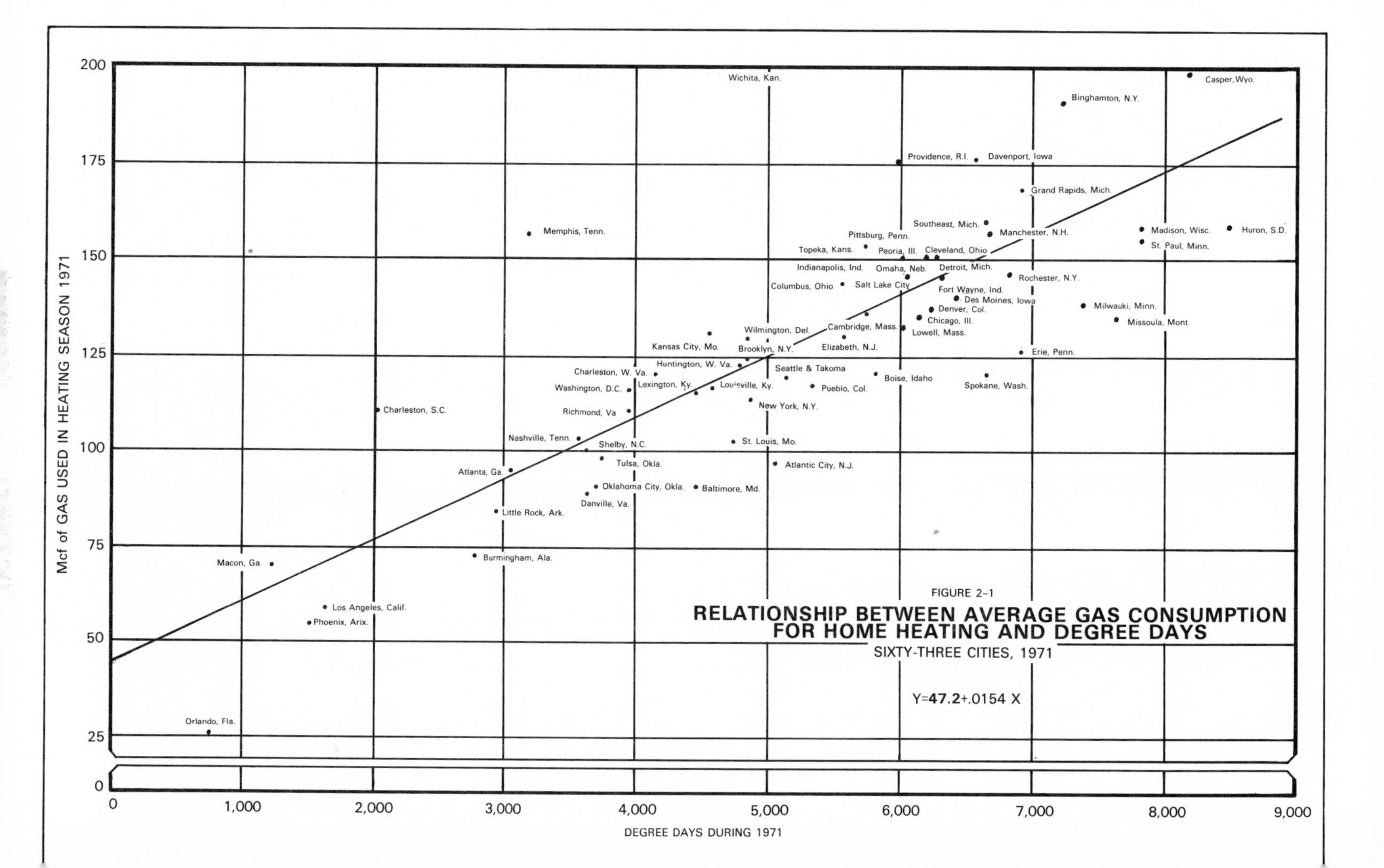

FIGURE 2-1

RELATIONSHIP BETWEEN AVERAGE GAS CONSUMPTION FOR HOME HEATING AND DEGREE DAYS

SIXTY-THREE CITIES, 1971

$Y = 47.2 + .0154\,X$

It must be stressed that the figures appearing in the tables on residential use (space heating as well as baseloads) represent monthly bills for specified volumes of energy. They are not the same as an average bill derived by dividing a utility's total revenue from residential customers by the number of customers. In the case of electricity, this latter measure would include the all-electric home in the same aggregate with a household using electricity for only lights and refrigeration. The resulting average expenditure figure would not represent a typical bill for either type of user because of the small proportion of residences using electricity for home heating. Even in 1970, in 70 per cent of the SMSAs included in this study, the proportion of homes heated with electric energy was under 10 per cent.

Residential Air Conditioning

The monthly bills appearing on the residential airconditioning table cover the costs of central airconditioning only. While figures from the 1970 Census of Housing indicate that in twenty out of the twenty-four SMSAs studied more dwelling places used window units than central systems, central systems have been priced in order to have comparable data for both gas and electricity.

Volumes representing the appropriate level of use were determined for each SMSA, and the cost was estimated according to the rates in the AGA *Rate Service* and the FPC *National Electric Rate Book* in the same manner as the baseloads and space heating components. The volume of energy consumed for airconditioning is the product of the following four variables: the size of the unit in use (in tons), the enrergy consumption per ton, the seasonal use expressed in hours, and the portion of seasonal use allocated to July. Both the seasonal use and the July ratio vary regionally. Data were obtained from the American Gas Association, the Edison Electric Institute, selected operating gas and electric utilities, and internal studies.

Commercial

The monthly energy costs for two different sizes of commercial establishments have been presented in this study. A small retail store would represent one level of use, while the requirements of a department store (or a large apartment house) would be typical of large customers.

For the small establishment, either gas or fuel oil would be selected as space heating fuel, while electricity would be used for lighting and other purposes for which there is no adequate power substitute. The volume of gas representing typical monthly use is 70 Mcf. This figure was obtained from data presented in *Gas Facts*[a], where both the volume

[a]Publication of the American Gas Association.

of natural gas sales by utilities to commercial customers and the number of such customers have been tabulated. Since no degree day adjustments have been made, and the same volume of use has been held constant for the small commercial category in all the SMSAs, some variation in the physical size of the store is implied.

On the basis of the Federal Power Commission publication *Typical Electric Bills,* supplemented by discussions with representatives of the utility industry, the level of electricity use selected as typical for this group of customers was 2,000 kilowatt hours per month (i.e., 250 hours of 8 kilowatts demand).

The same research and consultation established 5,000 Mcf per month as an appropriate measure of the volume of gas consumed by the large commercial user and 200,000 Kwh of electricity (333 hours of 600 Kw demand).

The fuel oil used by a small commercial establishment of the size contemplated in this study will, in all probability, be a #2 fuel oil. The quantity of #2 fuel oil consumed by small commercial establishments was derived by converting the volume used for gas to an equivalent volume of fuel oil, which was then raised 5 per cent to adjust for differences in the thermal efficiency of the two fuels. This adjustment factor is the same as that used in residential oil heat. The monthly consumption of #2 fuel oil for small commercial establishments in this study amounts to 530 gallons, a quantity, that for pricing purposes indicates a tankwagon delivery.

Large commercial establishments may use either #2 or #6 fuel oil. The quantities of these fuels used by large commercial establishments have been assumed to be equivalent to the gas consumption for this type of facility. Since the Btu's in a gallon of #2 differ from that in a gallon of #6 fuel oil, the quantity of #2 which is equivalent to the 5,000 Mcf of gas differs from that of #6. For #2 fuel oil the monthly quantity used in the study is 36,000 gallons, and for #6 fuel oil the quantity is about 34,000 gallons or 800 barrels. These quantities mean, for pricing purposes, truck transport deliveries.

Industrial

As the statistics in the *U.S. Census of Manufacturers* clearly portray, there is a wide variation in the energy needs of industrial establishments, where it is used not merely for space heating and lighting, but to power manufacturing machinery, run automatic controls, fuel boilers, etc. The monthly volumes selected have been 45,000 Mcf for gas, and 4,000,000 Kwh for electricity (i.e., 500 hours use of 8,000 Kw demand).[a]

[a]Also assumed for electricity are a 90.9 per cent power factor and a sub-transmission voltage delivery between 20-70 kilovolts to customers who provided their own transformer(s) to decrease the voltage.

Gas costs are presented under two sets of rate schedules, firm and interruptible. Firm rate schedules are usually used by industrial concerns having no facilities for using alternate fuels. Firm sales are essential for those industrial processes in which gas is needed for its intrinsic qualities above and beyond its Btu content, and in which other fuels would produce inferior results. Firm gas thus commands a premium price. Interruptible sales are subject to curtailment; customers must therefore maintain alternate fuel-burning facilities. Much of this gas is used in boilers.

A large industrial plant with fuel requirements equivalent to 45,000 Mcf per month, as was used in gas pricing, would need approximately 300,000 gallons per month of #2 fuel oil or about 7,200 barrels[a] per month of #6 fuel oil. For each of these two types of fuel oil, nothing less than truck transport loads would suffice, and prices were determined accordingly.

In the coal industry, an industrial user is classified as one that can take "on track" or "dockside" delivery. This definition was used in determining the steam coal prices to the industrial sector. Volumetrically, this would be equivalent to a minimum volume of 65 tons per shipment (i.e., the average capacity of a hopper car); the maximum volume would vary depending on the number of carloads or barges required by the user per shipment.

E. SOURCES OF RETAIL PRICE DATA

The monthly gas and electric bills for the retail sectors described above were calculated from actual rate schedules issued by the utilities.[b] These are usually priced for incremental volumes, with the unit cost declining as the volume of use increases. To the basic bills have been added state and local taxes, where they are in effect, as well as adjustments added by the utilities to cover increases in their cost of purchased fuel. Other applicable adjustments or discounts are noted in the footnotes accompanying each table.

This study has relied on published data for the bulk of its research material. However, for the later periods under consideration, it has been necessary to supplement this with more recent information obtained directly from the appropriate utility.

[a]A barrel of fuel oil is equal to 42 gallons.

[b]Gas rates are published by the American Gas Association in its *Rate Service*. Electric rates are available in the *National Electric Rate Book* issued by the Federal Power Commission.

Residential tankwagon prices of #2 fuel oil were based largely on data from the U.S. Bureau of Labor Statistics' Report *Retail Prices and Indexes of Fuels and Electricity.* This price information was supplemented by figures from *Fuel Oil and Oil Heat,* as well as from various oil companies. These sources, together with data from *The Oil Daily* also were utilized in the estimated monthly prices and bills for #2 fuel oil for small commercial establishments.

Prices of #2 fuel oil and #6 fuel oil for large commercial and industrial installations and service station prices of motor gasoline were obtained largely from *Platt's Oilgram Price Service.* Information from the *Oil and Gas Journal, Fuel Oil and Oil Heat,* and from various oil companies was also used. To the fuel oil prices, which were largely postings F.O.B. refinery or terminal, were added estimated delivery charges derived from information supplied by oil companies, the Interstate Commerce Commission, the Bulk Carrier Conference, Inc., and the U.S. Department of Commerce. In addition, sales taxes were included for pertinent SMSAs on the basis of data from the Commerce Clearing House and the American Petroleum Institute.

There is no publicly available source of information providing an historical series on industrial steam coal prices. The source of the data for industrial prices in this study was based in part on a private survey of coal producers most likely to supply the SMSAs and industrial users located in or around the SMSAs. Where there were gaps in the data time series or where only partial information was supplied, the data reported by the United States Bureau of Mines on average mine prices was used to develop an F.O.B. mine price series. Transportation costs were estimated from trends reported by the Bureau of Mines and data provided by questionnaire respondents.

Steam-electric generating plants of electric utilities are also retail purchasers of fuel. Data giving the prices paid by them for coal, oil, and gas are published each year by the National Coal Association in *Steam-Electric Plant Factors* and are based on annual reports (Forms No. 1, 1-M and 12) filed with the FPC. Unit costs have been presented on both a volume (i.e., per ton, Mcf or barrel) and a Btu basis. Coal and oil prices have been calculated "as burned" rather than "F.O.B." in order to be consistent with gas deliveries, which are presented only on an "as burned" basis.

The fuel prices for electric utilities are weighted averages of the prices paid by all of the generating plants in a utility's power system, not merely that paid by the facility closest to the SMSA being serviced, an approach necessitated by the utility's interconnecting grids.

F. WHOLESALE PRICE DEFINITIONS AND SOURCES

Wholesale prices are defined in this study as the price paid by the retailer to his supplier. Under this definition, the wholesale prices of some forms of the energy may be the price or cost of the basic producer of the energy. In other instances, one or more functions or middlemen may exist between the basic producer of the particular energy at its source and the ultimate sale to the retailer.

In coal sales to industrial users, the producer generally performs both the wholesale and the retail functions, selling the coal to the consumer. Most electric utilities not only produce the electricity, but also transmit and distribute it to the consumer with no wholesale price existing *per se*.

On the other hand, the gas sold by the utilities in most of the nation's SMSAs is purchased from the large interstate pipelines. The wholesale gas price in this study is the price the utility pays the pipeline at the "city gate."

The oil industry is more complex than the other energy industries because the basic crude oil requires refining before it is consumed. The sale of the refined oils in some markets may be direct from the refineries to the consumers, and consequently the refinery price may be both the wholesale and the retail price. On the other hand, in other markets, such as fuel oil for househeating and gasoline, one or more middlemen may occur between the refiner and the ultimate consumer, with distinct wholesale and retail prices existing.

GAS WHOLESALE PRICES

The gas wholesale or city gate price for each SMSA is the weighted average of the gas purchased by the distributor under all rate schedules including contract demand, interruptible, peaking, and storage, among others. The data are taken from the sales for resale tabulations appearing in the Form 2 *Annual Reports by Interstate Pipelines to the FPC*. In many instances, the local gas utility is served by more than one interstate supplier, and all supplies are included in the price. Where the bulk of or all of the supply was purchased from local intrastate sources, which do not report to the FPC, it was necessary to obtain figures from the distributor.

ELECTRIC WHOLESALE PRICE

Most electric utilities not only produce but also sell to the ultimate consumer, with no wholesale price existing as such[a]; therefore, the

[a] While utilities do purchase and interchange some electricity from each other, it is usually for such purposes as covering peak demands, emergencies, etc., and the pattern of such purchases is irregular over time.

wholesale price in this study has been imputed, based on the estimated cost of generation. The wholesale price is thus an estimated price or cost at the bus bar. It has been derived from data reported to the FPC by electric utilities in their annual reports.[a]

The procedure used in developing the imputed price has been to add on an estimate of fixed charges to the actually reported production expenses and to divide this sum by the total reported net electricity generated, resulting in a cost per kilowatt hour. The fixed charges associated with generation plants consist of five major elements—the cost of money, depreciation, interim relplacements, insurance, and taxes. These are related to the total capital cost or gross investment in the electric plant in service and are expressed by the FPC's Bureau of Power on a levelized basis as a percentage of this total investment. This percentage of total investment varies with ownership categories. For privately owned utilities used in this study it is 14 per cent; for publicly owned, 7.5 per cent[b]. The appropriate percentage was applied to the value of steam, nuclear, hydraulic, and other plant in service for each utility to derive the fixed charge estimate.

Two exceptions to the above procedure are Nashville Electric and Seattle Department of Lighting. The Tennessee Valley Authority functions as a wholesaler of electricity, generating it and selling it to local distributors throughout the area. The Seattle Department of Lighting, in addition to generating a portion of its requirements, purchases power from the Bonneville Power Administration. In these two instances the prices are actual wholesale prices rather than imputed.

OIL PRODUCTS WHOLESALE PRICES

The source of most of the wholesale prices for fuel oil (the reseller prices), and of the prices paid by gasoline service station dealers (the dealer tankwagon price) used in the study was *Platt's Oilgram Price Service.* Additional wholesale price information was obtained from *Fuel Oil and Oil Heat* and from various oil companies.

COAL WHOLESALE PRICES

Wholesale prices for steam coal were defined for the purpose of this study to be the F.O.B. mine price. These data were developed from questionnaire responses from coal producers and users and from U.S. Bureau of Mines data.

[a] Form 1 for investor-owned companies; Form 1-M for the government utilities.

[b] See *Hydroelectric Power Evaluation,* FPC P-35, 1968 and *Hydroelectric Power Evaluation Supplement No. 1,* FPC P-38, 1969. The application of different percentages for the two types of utilities stems from differences in their tax liabilities and in their cost of capital.

Chapter Three

Trends and Patterns in Energy Prices Retail Energy Prices

Chapter Five presents an overview of trends and patterns in the prices of the various major forms of energy to the different types of users in the twenty-four SMSAs covered by the study. While summary tables showing prices for selected SMSAs in the various Census regions are included in the text, for additional details see Tables 1 through 8 in the Appendix.

National retail energy prices (prices to the ultimate consumer) were relatively stable during most of the 1960's. The cost of electricity tended to show small decreases, while gas prices registered only modest changes, with decreases in rates in some SMSAs balancing gains in others. Heavy fuel oil prices generally were little changed. Distillate fuel oil rose noticeably beginning in 1966, with coal following the same pattern. In the latter part of the decade, pump prices of gasoline also posted gains, but frequent and widespread price wars produced a very erratic price pattern.

At the end of the 1960's, retail energy prices began a general upward movement that has continued to the present time. The largest increases have been in coal and heavy fuel oil, particularly in areas where the use of very low-sulfur oil has been required by government regulation. The downward trend in electricity prices has been reversed, although in general the increase in the price of electricity has been smaller than that of most other forms of energy.

A. RESIDENTIAL GAS AND ELECTRIC BASELOAD

The monthly bill for the non-substitutable electric baseload[a] of residential structures, covering such uses as lighting, refrigeration, and small ap-

[a]The electric baseload rate, both substitutable and non-substitutable, is based on homes without electric heat, while the gas baseload rate is for homes without gas heat.

pliances, has been rising in most SMSAs throughout the United States during the past four years (see Table 3-1). In the preceding nine years, from 1960 to 1969, it showed a varied pattern, with the monthly bill up, generally moderately, in some SMSAs and down, usually moderately, in others. The increases during the past four years, however, generally have been appreciably larger than the declines some of the SMSAs had experienced in the previous nine years. As of January 1973, New York City had the highest non-substitutable baseload monthly bill, followed by Chicago, while Nashville and Seattle (both cities served by federal power projects) had the lowest.

For the substitutable baseload such as water heaters, ranges, and clothes dryers, in which either gas or electricity may be used, the monthly bill for electricity in most SMSAs declined from 1960 to 1969, and then rose markedly during the past four years. The gas substitutable baseload monthly bill showed a varied pattern among the different SMSAs during the 1960's, rising in some SMSAs and declining in others; but since January 1969 the bill for gas rose in practically all SMSAs. Gas has continued to remain cheaper than electricity for substitutable baseload use in practically all regions of the nation.

B. RESIDENTIAL HOUSE HEATING

The average monthly bill for space heating of a home with gas, as well as that for electricity, declined from January 1960 to January 1969 in most SMSAs of the nation, with particularly sharp declines observed in some SMSAs for electricity (see Table 3-2). This declining trend has been reversed in the past four years as both gas and electric rates have gone up markedly. Gains ranging up to 30 per cent for gas and up to 50 per cent for electricity have occurred in the eastern half of the nation.

Space heating costs for a home using fuel oil, unlike the gas- or electric-heated house, increased noticeably between 1960 and 1969, but most of the increase came after 1965. The rising trend in the fuel oil space heating cost has continued to date, although its relative increase from 1969 to 1972 tended to be less than for gas and electricity.

Electric space heating costs, even allowing for greater thermal efficiency, are substantially higher than for either gas or oil throughout the nation's SMSAs. Gas costs, for the most part, continue to be significantly lower than fuel oil costs, except in such northeastern SMSAs as Boston and New York City.

Census Region / Census Region SMSA	January 1960 Non-Substitutable Electric	January 1960 Substitutable Electric	January 1960 Substitutable Gas	January 1969 Non-Substitutable Electric	January 1969 Substitutable Electric	January 1969 Substitutable Gas	January 1973 Non-Substitutable Electric	January 1973 Substitutable Electric	January 1973 Substitutable Gas
New England									
Boston, Mass.	$14.29	$24.10	$10.36	$13.40	$15.60	$10.30	$14.50	$17.78	$14.09
Middle Atlantic									
New York, N.Y.	16.08	17.54	9.41	21.08	15.57	9.06	17.11	21.91	10.44
South Atlantic									
Charlotte, N.C.	9.64	16.48	8.01	9.54	11.69	7.10	11.46	14.06	7.49
East North Central									
Chicago, Ill.	11.69	17.06	6.03	11.41	15.16	5.94	13.79	19.36	6.84
East South Central									
Nashville, Tenn.	6.69	3.30	6.32	7.00	4.25	6.32	8.99	7.95	7.39
West North Central									
Minneapolis-St. Paul, Minn.	11.67	18.27	5.96	10.45	16.73	6.10	12.64	14.42	7.73
West South Central									
Houston, Tex.	8.13	10.80	4.41	9.56	8.69	6.12	10.58	9.20	6.62
Mountain									
Denver, Colo.	9.73	16.32	2.85	10.89	15.90	3.61	11.26	17.94	4.00
Pacific									
Los Angeles, Calif.[a]	8.03	10.18	5.87	8.16	10.36	5.64	11.43	14.66	6.96
[b]	8.97	10.15	—	8.72	10.94	—	12.35	13.44	—
Seattle Wash.[a]	5.74	6.22		5.77	6.19		6.49	6.56	
[b]	8.03	6.45	12.95	8.83	7.82	12.78	9.29	7.82	9.57c

[a] For electric baseloads—government utility.
[b] For electric baseloads—investor-owned utility.
[c] New Rate Schedule.

Note: Non-substitutable baseload assumes no electric substitutable baseload.
Substitutable baseload—electric—assumes no electric space heating.
Substitutable baseload—gas—assumes no gas space heating.

Source: See Tables 1-a through 1-x in the Appendix.

Table 3-2. Residential Monthly Bill for Gas, Electric and Fuel Oil Space Heating Use (In Dollars)

Census Region SMSA	January 1960			January 1969			January 1973		
	Gas	*Electric*	*Oil*	*Gas*	*Electric*	*Oil*	*Gas*	*Electric*	*Oil*
New England									
Boston, Mass.	33.02	147.01	25.59	31.81	78.08	30.29	41.72	91.36	34.56
Middle Atlantic									
New York, N.Y.	27.16	94.97	22.63	25.04	58.60	26.69	30.37	87.13	30.59
South Atlantic									
Charlotte, N.C.	19.49	58.42	19.24	16.83	39.56	20.94	18.65	47.57	21.96
East North Central									
Chicago, Ill.	24.41	93.68	25.55	22.86	58.47	28.93	28.59	73.06	32.06
East South Central									
Nashville, Tenn.	20.96	35.78	—	20.96	35.99	—	26.41	44.78	—
West North Central									
Minneapolis-St. Paul, Minn.	24.48	99.56	29.32	23.11	81.98	32.21	26.26	86.96	34.99
West South Central									
Houston, Tex.	6.60	20.98	—	8.11	20.66	—	9.11	21.92	—
Mountain									
Denver, Colo.	6.76	100.90	—	10.63	62.91	—	11.42	79.28	—
Pacific									
Los Angeles,	8.63	29.07[a]	—	8.46	29.36[a]	—	8.91	36.31[a]	—
Calif.	—	31.06[b]	—	—	31.67[b]	—	—	31.92[b]	—
Seattle	30.30	38.85[a]	26.62	24.92	39.04[a]	30.54	27.41	40.40[a]	33.92
Wash.	—	46.18[b]		—	52.57[b]	—	—	55.68[b]	

[a] Government utility.

[b] Investor-owned utility.

Note: 1. For electric space heat assumes an electric substitutable base load and for gas heat assumes a gas substitutable base load.
The monthly bills for electricity and gas are based on the rates in effect on January of the respective years. The monthly bill for oil is

Note: 2. based on an average of No. 2 fuel oil prices in January and July of the given year and January in the succeeding year.

Source: See Tables 1-a to 1-x in Appendix.

C. RESIDENTIAL AIR CONDITIONING

The average monthly bill for both gas and electricity for central air conditioning declined in the majority of the 24 SMSAs in this energy price study during the period from 1960 to 1969 (see Table 3-3). Particularly sharp declines for gas were noted in Miami, where there was a change in the rate schedules, and also in Chicago, while declines approaching 20 per cent occurred in Atlanta, Los Angeles, and Seattle. Meanwhile, declines of almost 30 per cent or more occurred in electric air conditioning rates for central air conditioning in Boston, Charlotte, Denver, and Salt Lake City. While the average monthly air conditioning bills for gas and electricity in a preponderance of SMSAs were lower in 1969 than in 1960, in a significant number of SMSAs the bills rose. In most cases, however, the increases were small to modest.

Gas and electricity rates for residential central air conditioning have also participated in the general rise in energy prices of the past several years. The average monthly bill for electricity use in air conditioning rose from 1969 to 1972 in all but one of the twenty-four SMSAs in this study, with the one exception being San Francisco.

Table 3-3. Residential Monthly Bill for Gas and Electric Air Conditioning (In Dollars)

	July 1960		*July 1969*		*July 1972*	
Census Region SMSA	*Gas*	*Electric*	*Gas*	*Electric*	*Gas*	*Electric*
New England						
Boston, Mass.	16.19	19.57	15.76	10.90	14.43	13.32
Middle Atlantic						
New York, N.Y.	18.32	16.77	17.11	18.98	23.67	24.20
South Atlantic						
Charlotte, N.C.	21.85	36.57	21.85	25.60	25.01	30.62
East North Central						
Chicago, Ill.	17.22	22.52	11.48	19.68	12.73	24.99
East South Central						
Nashville, Tenn.	14.77	13.23	14.77	13.41	19.13	17.11
West North Central						
Minneapolis-St. Paul, Minn.	13.54	12.58	13.36	12.96	14.52	13.34
West South Central						
Houston, Tex.	16.29	38.96	16.91	38.71	20.17	38.96
Mountain						
Denver, Colo.	5.19	16.85	6.71	10.50	7.43	13.13
Pacific						
Los Angeles, Calif.	5.30	4.26[a]	4.32	4.43[a]	4.84	5.89[a]
	—	4.65[b]	—	4.74[b]	—	5.76[b]

[a] Government utility.
[b] Investor-owned utility.
Source: See Table 2 in Appendix.

All but a few of the SMSAs also showed increases in gas air conditioning rates during this recent period. Moreover, many of the increases in both electricity and gas rates were substantial. Despite these recent increases, however, in about one-third of the twenty-four SMSAs, the monthly bill for gas and electricity for central air conditioning was lower in July 1972 than in July 1960.

The monthly gas and electric bills for central air conditioning are strongly affected by climatic conditions as well as by rates. As would be expected, the SMSAs with the highest bills for these fuels are largely southern SMSAs. Those with the lowest bills are on the West Coast.

The monthly gas bills for central air conditioning in 1972 were lower than the comparable electric bills in more than three-fourths of the twenty-four SMSAs in the study. The largest disparity between gas and the electric monthly air conditioning bills was in the SMSAs in the Mountain and the West South Central regions. SMSAs in which the monthly electric air conditioning bill was lower than the gas bill are scattered, but include Nashville, which is supplied electricity by a federal power authority.

D. COMMERCIAL AND INDUSTRIAL USE

The cost of heating small commercial establishments in national SMSAs has followed a pattern similar to that of residential heating. The types of fuels used are generally the same, and the changes in the prices of the different fuels tend to affect the heating costs of small commercial establishments similarly to those of residential establishments.

The fuel cost per month for heating small commercial establishments with the same requirements have been the highest in New England, followed by the Middle Atlantic and Pacific Northwest areas (see Table 3-4). Meanwhile, the lowest have been in the Mountain and West South Central areas.

Gas prices have moved up nationally during the past several years, but a similar rise has occurred in distillate fuel oils. In 1972, the average monthly bill of commercial establishments with the same requirements was lower for gas than for #2 fuel oil in most SMSAs except those on the East Coast and the Pacific Northwest.

Electricity rates for small commercial establishments, which had shown little change or even declines from 1960 to 1969, have shown widespread increases during the past few years (see Table 3-5). The largest increase has been in New York City, where the monthly bill rose 46 per cent from January 1969 to January 1973. Relatively large gains

Table 3-4. Gas and No. 2 Fuel Oil—Average Monthly Bill for Heating Small Commercial Establishments[a] (In Dollars)

Census Region SMSA	Gas January 1960	Gas July 1960	No.2 Fuel Oil 1960	Gas January 1969	Gas July 1969	No.2 Fuel Oil 1969	Gas January 1972	Gas July 1972	No.2 Fuel Oil 1972
New England									
Boston, Mass.	$153.25	$153.25	$77.90	$152.33	$158.46	$ 92.80	$180.36	$180.36	$106.50
Middle Atlantic									
New York, N.Y.	117.85	117.85	77.90	119.46	120.80	92.20	135.46	139.21	106.00
South Atlantic									
Charlotte, N.C.	92.94	92.94	75.80	88.52	88.52	82.70	88.52	95.90	86.90
East North Central									
Chicago, Ill.	76.70	76.62	75.80	62.00	60.53	85.90	78.64	76.60	95.40
East South Central									
Nashville, Tenn.	87.48	97.30	—	87.48	87.48	—	97.39	104.01	—
West North Central									
Minneapolis-St. Paul, Minn.	64.92	64.92	77.90	62.62	66.48	85.90	72.76	72.76	93.30
West South Central									
Houston, Tex.	44.76	44.76	—	52.32	52.32	—	60.15	60.15	—
Mountain									
Denver, Colo.	22.72	25.31	—	34.55	34.50	—	38.44	38.44	—
Pacific									
Seattle, Wash.	131.56	131.56	89.50	117.46	117.46	103.40	117.90	117.90	115.00

[a]Note: Requirements assumed to be the same for all cities.

Source: See Table 4-a in Appendix.

have also occurred in TVA-served Nashville, as well as in Chicago. The lowest rates, however, continue to prevail in Nashville, while the highest are in New York City, followed by Boston.

The trends in energy costs for large commercial establishments conformed with those evident for residential and small commercial users, generally showing little change during most of the 1960s and a significant rise since 1969. (See Tables 3-6, 3-7, 3-8, and 3-9.) The relative increases for gas and electricity since 1969, however, have been markedly higher for large commercial establishments than for small commercial or residential users. Distillate fuel oil costs for large commercial use, however, did not increase as much as for smaller consumers from 1969 to 1972.

Table 3-5. Electricity—Average Monthly Bill of Small Commercial Establishments (In Dollars)

Census Region	*1960*		*1969*		*1972*		*1973*
SMSA	*January*	*July*	*January*	*July*	*January*	*July*	*January*
New England							
Boston, Mass.	86.07	85.78	82.00	82.00	90.10	89.27	87.44
Middle Atlantic							
New York, N.Y.	69.46	68.86	67.79	68.06	94.95	98.64	99.22
South Atlantic							
Charlotte, N.C.	49.44	49.44	48.41	48.41	49.48	58.21	58.21
East North Central							
Chicago, Ill.	78.14	78.14	73.69	74.20	85.04	85.63	85.36
East South Central							
Nashville, Tenn.	26.78	26.78	27.89	31.11	37.18	37.18	37.18
West North Central							
Minneapolis-St. Paul, Minn.	77.75	77.75	59.70	59.70	67.97	68.23	68.75
West South Central							
Houston, Tex.	46.90	47.40	47.45	47.45	48.63	48.63	49.58
Mountain							
Denver, Colo.	60.95	60.95	67.84	67.84	67.84	71.38	71.38
Pacific							
San Fran., Calif.	61.83	61.83	61.97	61.97	70.69	70.69	70.69

Source: See Table 4-d in Appendix.

Table 3-6. Electricity—Price to Large Commercial Users (Cents Per Kwh)

Census Region SMSA	*1960 January*	*1960 July*	*1969 January*	*1969 July*	*1972 January*	*1972 July*	*1973 January*
New England							
Boston, Mass.	2.07	2.06	1.91	1.91	2.33	2.29	2.20
Middle Atlantic							
New York, N.Y.	1.97	1.94	2.19	2.50	3.28	3.46	3.49
South Atlantic							
Charlotte, N.C.	1.11	1.11	1.10	1.10	1.15	1.32	1.32
East North Central							
Chicago, Ill.	1.57	1.57	1.61	1.62	2.03	2.06	2.05
East South Central							
Nashville, Tenn.	0.73	0.73	0.77	0.84	1.10	1.10	1.10
West North Central							
Minneapolis-St. Paul, Minn.	1.78	1.78	1.60	1.79	1.86	1.87	1.88
West South Central							
Houston, Tex.	1.08	1.10	1.20	1.20	1.34	1.33	1.37
Mountain							
Denver, Colo.	1.33	1.33	1.49	1.49	1.62	1.67	1.63
Pacific							
San Fran., Calif.	1.34	1.34	1.37	1.37	1.53	1.53	1.53

Source: See Table 5-d in Appendix.

Table 3-7. Gas—Price to Large Commercial Users (Cents Per Mcf)

Census Region SMSA	*1960 January*	*1960 July*	*1969 January*	*1969 July*	*1972 January*	*1972 July*	*1973 January*
New England							
Boston, Mass.	138.1	138.1	136.8	137.7	181.7	181.7	201.0
Middle Atlantic							
New York, N.Y.	137.3	137.3	131.7	133.6	154.1	159.7	159.8
South Atlantic							
Charlotte, N.C.	123.7	123.7	72.1	72.1	72.1	82.7	82.7
East North Central							
Chicago, Ill.	71.0	70.9	64.4	62.3	88.2	83.9	89.5
East South Central							
Nashville, Tenn.	74.1	77.4	74.1	74.1	88.2	94.2	98.3
West North Central							
Minneapolis-St. Paul, Minn.	70.0	70.0	71.3	76.2	92.0	92.0	92.0
West South Central							
Houston, Tex.	45.2	45.2	47.3	47.3	63.3	63.3	63.3
Mountain							
Denver, Colo.	30.0	33.9	46.6	46.6	49.2	49.2	49.2
Pacific							
Seattle, Wash.	76.3	76.3	77.0	77.0	77.6	77.6	90.5

Source: See Table 5-d in Appendix.

Table 3-8. No. 2 Fuel Oil—Delivered Prices to Large Commercial Users (Cents Per Gallon)

Census Region	*1960*		*1969*		*1972*		*1973*
SMSA	*January*	*July*	*January*	*July*	*January*	*July*	*January*
New England							
Boston, Mass.	10.4	9.6	11.5	12.0	12.5	12.6	12.6
Middle Atlantic							
New York, N.Y.	11.1	10.0	12.0	12.1	13.1	13.2	13.2
South Atlantic							
Charlotte, N.C.	11.4	10.6	12.9	12.9	12.7	12.8	12.9
East North Central							
Chicago, Ill.	10.4	9.8	11.5	11.0	12.2	12.3	12.3
East South Central							
Nashville, Tenn.	—	—	—	—	—	—	—
West North Central							
Minneapolis-St. Paul, Minn.	11.2	9.9	11.7	11.5	11.9	11.9	12.1
West South Central							
Houston, Tex.	—	—	—	—	—	—	—
Mountain							
Denver, Colo.	—	—	—	—	—	—	—
Pacific							
Seattle, Wash.	11.5	11.5	12.2	12.6	14.1	14.1	14.1

Source: See Table 5-d in Appendix.

Table 3-9. No. 6 Fuel Oil—Delivered Prices to Large Commercial Users[1] (Dollars Per Barrel)

Census Region	*1960*		*1969*		*1972*		*1973*
SMSA	*January*	*July*	*January*	*July*	*January*	*July*	*January*
New England							
Boston, Mass.	2.77	2.92	2.57	2.15	5.16[a]*	4.70[a]*	5.20[a]*
Middle Atlantic							
New York, N.Y.	2.73	2.97	2.76[b]	2.79[b]	5.74[c]	5.16[c]	5.70[c]
South Atlantic							
Charlotte, N.C.	—	—	—	—	—	—	—
East North Central							
Chicago, Ill.	3.58[d]	3.54[d]	3.80[e]	3.69[e]	5.16[e]	5.17[e]	5.17[e]
East South Central							
Nashville, Tenn.	—	—	—	—	—	—	—
West North Central							
Minneapolis-St. Paul, Minn.	3.74	3.57	3.91	4.24	4.74	4.76	4.91
West South Central							
Houston, Tex.	—	—	—	—	—	—	—
Mountain							
Denver, Colo.	—	—	—	—	—	—	—
Pacific							
Seattle, Wash.	3.12[b]	3.12[b]	3.29[b]	3.30[b]	4.79[b]	4.85[b]	4.82[b]

*Estimated.

[a]0.5% sulfur. [b]2.0% sulfur max. [c]0.3% sulfur max. [d]Low sulfur. [e]1.0% sulfur max.

[1]Sulfur content is regular or no sulfur guarantee unless specified.

Source: See Table 5-c in Appendix.

While energy prices paid by large industrial users (see Tables 3-11, 3-12, 3-13, 3-14, and 3-15) have been lower than those of large commercial establishments, the trend in prices of the two types of users has tended to parallel each other. Accordingly, for gas and electricity, the increases in recent years have been relatively greater than for residential and small commercial users, the same situation as has prevailed for large commercial users. For fuel oils, however, taxes have tended to push up commercial costs more than industrial costs. Coal prices, which affect largely industrial plants and electric utilities, have shown greater increases since 1969 than any of the other forms of energy, and this has been a factor pushing up industrial fuel costs.

Industrial users are typically more sensitive to fuel costs than other types of users. In view of this situation, Table 3-10 is presented to show the cost of various fuels for large industrial consumers in terms of cents per million Btu for selected SMSAs in different census regions. Costs of coal and fuel oil have been omitted where their use is relatively small.

Table 3-10. Comparison of Primary Fuel Costs to Large Industrial Users (cents per million Btu)

	January 1960				January 1969				January 1973			
		Fuel Oil				Fuel Oil				Fuel Oil		
Census Region SMSA	*Gas*	*No2*	*No.6*	*Coal*	*Gas*	*No.2*	*No.6*	*Coal*	*Gas*	*No.2*	*No.6*	*Coal*
New England												
Boston, Mass.	109.7	75.0	44.3	—	134.0	83.6	40.9	—	197.7	90.9	84.9	—
Middle Atlantic												
New York, N.Y.	122.5	77.9	42.4	42.9[a]	121.3	82.2	42.1	44.8[a]	146.4	89.4	87.0	73.4[b]
South Atlantic												
Charlotte, N.C.	121.1	80.8	—	32.7[a]	70.6	91.2	—	42.8[a]	81.0	88.7	—	72.0[c]
East North Central												
Chicago, Ill.	63.8	75.0	57.8	32.1[a]	56.3	82.9	59.9	36.7[a]	80.7	88.7	84.4	50.2[c]
East South Central												
Nashville, Tenn.	48.9	—	—	34.9[a]	49.3	—	—	44.8[a]	68.9	—	—	73.8[c]
West North Central												
Minneapolis-St. Paul, Minn.	68.2	80.8	59.8	57.1	67.1	82.2	60.6	63.9	76.2	82.9	73.2	96.9
West South Central												
Houston, Tex.	18.3[a]	—	—	—	19.4[a]	—	—	—	25.2[c]	—	—	—
Mountain												
Denver, Colo.	11.4	—	—	32.5[c]	23.0	—	—	31.5[a]	25.6	—	—	37.7[a]
Pacific												
Seattle, Wash.	67.7	82.9	49.9	40.7	65.3	88.0	52.6	53.5	81.6	101.7	77.1	77.4

[a] Annual averages

[b] Price for 1971; later price not available.

[c] Annual average for 1972.

Note: Costs for gas are based on rates for firm gas.

Source: Data based on Tables 5a-5d in Appendix.

Residual fuel oil has been the lowest-cost fuel for industrial use in the coastal parts of the nation, although the cost difference between fuel oil and other fuels has recently narrowed or disappeared, primarily because of sulfur regulations. Meanwhile, coal has been priced lower in the East North Central region and in some areas of the South. Gas has had a price advantage over other fuels in the West South Central region, in some Mountain Region SMSAs such as Denver, and in California.

The pattern of energy costs for large industrial users in the SMSAs reflects the importance of transportation costs. By and large, the lowest-cost fuel in a particular SMSA has been the one whose major sources of supply have been relatively close, or for which relatively low-cost water transportation could be used.

Table 3-11. Electricity—Price to Large Industrial Users (Cents Per Kwh)

Census Region	*1960*		*1969*		*1972*		*1973*
SMSA	*January*	*July*	*January*	*July*	*January*	*July*	*January*
New England							
Boston, Mass.	1.22	1.20	1.18	1.18	1.57	1.53	1.44
Middle Atlantic							
New York, N.Y.	1.48	1.45	1.63	1.64	2.35	2.54	2.57
South Atlantic							
Charlotte, N.C.	0.69	0.69	0.69	0.69	0.74	0.83	0.83
East North Central							
Chicago, Ill.	0.99	0.99	1.03	1.05	1.38	1.41	1.39
East South Central							
Nashville, Tenn.	0.49	0.49	0.53	0.56	0.77	0.77	0.77
West North Central							
Minneapolis-St. Paul, Minn.	0.98	0.98	0.93	1.06	1.16	1.18	1.20
West South Central							
Houston, Tex.	0.66	0.69	0.62	0.62	0.69	0.68	0.72
Mountain							
Denver, Colo.	0.79	0.79	0.87	0.87	0.93	0.95	0.94
Pacific							
San Fran., Calif.	0.85	0.85	0.85	0.85	0.94	0.94	0.94

Source: See Table 6-f in Appendix.

Table 3-12. Gas—Price of Firm Gas to Large Industrial Customers (Cents Per Mcf)

Census Region SMSA	*1960 January*	*1960 July*	*1969 January*	*1969 July*	*1972 January*	*1972 July*	*1973 January*
New England							
Boston, Mass.	109.7	109.7	134.0	134.1	179.4	179.4	198.7
Middle Atlantic							
New York, N.Y.	127.3	127.3	123.5	125.3	143.5	148.5	148.6
South Atlantic							
Charlotte, N.C.	121.1	121.1	70.6	70.6	70.6	81.0	81.0
East North Central							
Chicago, Ill.	64.0	63.9	58.3	56.1	82.0	77.7	83.4
East South Central							
Nashville, Tenn.	52.8	57.2	52.8	52.8	66.4	72.1	71.7
West North Central							
Minneapolis-St. Paul, Minn.	68.2	68.2	68.1	73.0	76.7	76.7	76.7
West South Central							
Houston, Tex.[a]	19.0		20.0		26.0		n,a.
Mountain							
Denver, Colo.	29.5	33.1	44.0	44.3	46.3	46.3	46.3
Pacific							
Seattle, Wash.	67.7	67.7	68.2	68.2	73.0	73.0	85.7

[a] Annual average calculated from gas deliveries divided by number of industrial customers.

Source: See Table 6-a in Appendix.

Table 3-13. No. 2 Fuel Oil—Delivered Prices to Large Industrial Users (Cents Per Gallon)

Census Region SMSA	*1960 January*	*1960 July*	*1969 January*	*1969 July*	*1972 January*	*1972 July*	*1973 January*
New England							
Boston, Mass.	10.4	9.6	11.6	11.6	12.5	12.6	12.6
Middle Atlantic							
New York, N.Y.	10.8	9.7	11.4	11.4	12.3	12.4	12.4
South Atlantic							
Charlotte, N.C.	11.2*	10.4*	12.7	12.7	12.3	12.3	12.3
East North Central							
Chicago, Ill.	10.4	9.8	11.5	11.0	12.2	12.3	12.3
East South Central							
Nashville, Tenn.	—	—	—	—	—	—	—
West North Central							
Minneapolis-St. Paul, Minn.	11.2	9.9	11.4	11.0	11.4	11.4	11.5
West South Central							
Houston, Tex.	—	—	—	—	—	—	—
Mountain							
Denver, Colo.	—	—	—	—	—	—	—
Pacific							
Seattle, Wash.	11.5*	11.5	12.2	12.6	14.1	14.1	14.1

*Estimated.

Source: See Table 6-B in Appendix.

Table 3-14. No. 6 Fuel Oil—Delivered Prices to Large Industrial Users[1] (Dollars Per Barrel)

Cencus Region SMSA	*1960 January*	*1960 July*	*1969 January*	*1969 July*	*1972 January*	*1972 July*	*1973 January*
New England							
Boston, Mass.	2.77	2.92	2.56	25.7	5.16[a]*	4.70[a]*	5.20[a]*
Middle Atlantic							
New York, N.Y.	2.65	2.88	2.63[b]	2.63[b]	5.36[c]	4.82[c]	5.33[c]
South Atlantic							
Charlotte, N.C.	—	—	—	—	—	—	—
East North Central							
Chicago, Ill.	3.58[d]	3.54[d]	3.67[e]	3.79[e]	5.16[e]	5.17[c]	5.17[c]
East North Central							
Nashville, Tenn.	—	—	—	—	—	—	—
West North Central							
Minneapolis-St. Paul, Minn.	3.74	3.57	3.79	3.80	4.56	4.58	4.58
West South Central							
Houston, Tex.	—	—	—	—	—	—	—
Mountain							
Denver, Colo.	—	—	—	—	—	—	—
Pacific							
Seattle, Wash.	3.12	3.12	3.29	3.30	4.79	4.81	4.82

*Estimated.

[a]0.5% sulfur max. [b]2.0% sulfur max. [c]0.3% sulfur max. [d]Low sulfur. [e]1.0% sulfur max.

[1]Sulfur content is regular or no sulfur guarantee unless specified.

Source: See Table 6-c in Appendix.

Table 3-15. Coal—Delivered Prices to Large Industrial Customers (Dollars Per Ton)

Census Region	*1960*		*1969*		*1972*		*1973*
SMSA	*July*	*Dec.*	*July*	*Dec.*	*July*	*Dec.*	*March*
New England							
Boston, Mass.	—	—	—	—	—	—	—
Middle Atlantic							
New York, N.Y.[a]	12.22		12.54		20.40[b]		—
South Atlantic							
Charlotte, N.C.[a]	9.18		12.27		20.42		21.39
East North Central							
Chicago, Ill.[a]	7.58		8.67		11.86		14.42
East South Central							
Nashville, Tenn.[a]	9.80		12.86		20.93		21.89
West North Central							
Minneapolis-St. Paul, Minn.	16.45	16.45	18.40	20.20	27.90	27.90	28.50
West South Central							
Houston, Tex.	—	—	—	—	—	—	—
Mountain							
Denver, Colo.[a]	6.92		6.72		8.03		8.43
Pacific							
Seattle, Wash.	10.18	10.42	13.35	14.42	17.79	19.36	19.45

[a] Annual averages.

[b] Price for 1971; 1972 price not available.

Source: See Table 6-d in Appendix.

E. FOSSIL FUELS USED BY ELECTRIC UTILITIES

Prices paid by electric utilities for all three major fossil fuels—coal, oil, and gas—(see Table 3–16) are available each year of the 1960's and early 1970's for only six of the twenty-four SMSAs of the study. This is because all three fuels have not been competitive alternatives for power plant use in most of the areas. Generally, one fuel or another has dominated power plant use in a particular area. Even in the six metropolitan centers in which the electric utilities have reported prices for all three fuels for each year of the study period, the fuels have not been of equal importance nor consistently used. For instance, gas may have had a competitive edge only when the utilities in some northern cities purchased it on an interruptible basis and burned other fuels when the gas was cut off.

Table 3–16. Prices of Fuel to Power Plants of Electric Utilities (Cents Per Million Btu)

Census Region		*1960*			*1969*			*1972*	
SMSA	*Coal*	*Oil*	*Gas*	*Coal*	*Oil*	*Gas*	*Coal*	*Oil*	*Gas*
New England									
Boston, Mass.	36.5	35.0	—	—	27.7	—	—	65.2	—
Middle Atlantic									
New York, N.Y.	36.3	35.0	38.7	39.8	36.9	37.1	71.0	65.0	57.7
South Atlantic									
Charlotte, N.C.	26.9	—	—	29.3	—	27.1	43.9	83.8	53.8
East North Central									
Chicago, Ill.	29.3	72.8	25.1	28.0	—	27.7	43.2	60.4	54.7
East South Central									
Nashville, Tenn.	18.7	—	—	19.7	—	20.5	29.9	—	n.a.
West North Central									
Minneapolis-St. Paul, Minn.	27.6	87.0	28.3	30.4	68.7	24.9	41.0	77.5	31.5
West South Central									
Houston, Tex.	—	—	18.7	—	—	19.8	—	43.3	21.3
Mountain									
Denver, Colo.	25.1	36.1	22.3	25.8	25.3	22.9	n.a.	n.a.	n.a.
Pacific									
San Fran. Calif.	—	32.6	35.1	—	29.5	31.8	—	48.2	38.4

N.A.—not available.

Source: See Table 7–a through 7–d in Appendix.

The oil prices reported by the utilities generally have been for residual fuel, although sometimes an indistinguishable amount of distillate fuel oil has been included. By and large, the price of distillate fuel oil has been too high for it to be used as the major fuel in a steam generating plant; it has, however, been used to some extent to aid coal combustion, to blend with heavy fuel oil, or for gas turbines. Some of the oil prices that appear very high in relation to the other fuels used by power plants in an area may apply largely to the relatively small portion of distillate fuel oil used by the power plant.

The fuel prices paid by the electric utilities generally have been significantly lower than the prices paid for the same fuel by other users. This phenomenon is due to the tremendous quantity of fuel the utilities purchase, which is frequently greatly in excess of that bought by other users. Another factor tending to produce lower fuel prices for utilities is that utilities use long-term contracts.

Fuel prices paid by the utilities tended to decline during the early 1960's, with coal prices turning up somewhat earlier than gas and oil prices. During the past few years, prices of all three fuels have risen markedly.

The geographic pattern of electric utility fuel prices is similar to that for other types of fuel users. The highest fuel costs by the utilities have been in the Northeast, and the lowest utility fuel costs have been in the West South Central region.

F. MOTOR GASOLINE

Throughout the 1960's and up until recent months service station prices of major brand gasoline have been subject to frequent and relatively widespread depressed pricing, which has tended to obscure and distort basic price trends. In fact, in some SMSAs the depressed pricing has been so frequent that the term "basic trend" may appear to be meaningless. Nevertheless, if allowance is made for the episodes of depressed prices, service station prices showed a generally sidewise trend during the first half of the 1960's, and then evidenced a persistent and marked upward trend since 1966 (see Table 3–17).

The variation in prices of retail gasoline among the nation's SMSAs is much smaller than for such energy sources as natural gas, coal, and electricity. (Distillate fuel oil prices, however, also show a relatively small variation). Despite the relatively small variation, gasoline prices in Houston tend to be lower than in other SMSAs, while prices in New York City tend to be the highest. The higher prices in New York City are due in part to higher taxes, which include state and local sales taxes and a lead tax.

Table 3-17. Service Station Prices* of Major Brand Regular Grade Gasoline (Cents Per Gallon)

Census Region SMSA	*1960 January*	*1960 July*	*1966 January*	*1966 July*	*1972 January*	*1972 July*	*1973 January*
New England							
Boston, Mass.	22.9	27.9	29.9	30.9	37.9	34.9	35.9
Middle Atlantic							
New York, N.Y.	33.9	31.9	32.9	32.9	41.9	41.9	41.9
South Atlantic							
Charlotte, N.C.	31.9	30.9	32.9	32.9	38.9	36.9	38.9
East North Central							
Chicago, Ill.	31.9	32.9	33.9	33.9	37.3	33.9	40.4
East South Central							
Nashville, Tenn.	30.9	30.9	31.9	33.9	35.9	35.9	35.9
West North Central							
Minneapolis-St. Paul, Minn.	29.9	31.4	31.9	32.9	38.9	36.9	36.9
West South Central							
Houston, Tex.	29.9	29.9	28.9	29.9	33.9	31.9	31.9
Mountain							
Denver, Colo.	27.9	33.9	29.9	32.9	35.9	36.9	34.9
Pacific							
Los Angeles, Calif.	29.9	29.9	28.9	32.9	34.9	34.9	37.9

*Includes excise and sales taxes. (See Table 9–d for taxes.)

Source: See Table 3 in Appendix.

G. WHOLESALE ENERGY PRICES

Movements in wholesale energy prices, as would be expected, have shown a similar pattern to retail prices. The imputed wholesale price of electricity registered little change during the 1960's, but has risen in most SMSAs in recent years, with dramatic increases occurring in the eastern half of the country (see Table 3–18). Wholesale prices of natural gas (the city gate price) and of heavy fuel oil were relatively stable from 1960 through 1968, but then had marked increases in the subsequent four years (see Table 3–19). The wholesale price of coal (the mine-mouth price), after little change from 1960 to 1965, edged up in the next four years and since 1969 has shown large increases. Meanwhile, the distillate fuel oil wholesale price and the dealer tankwagon price of gasoline followed a sidewise trend during the first half of the 1960's, but have recently been showing an upward movement (see Table 3–30).

The largest increases in wholesale energy prices in recent years have been in coal and heavy fuel oil in East Coast States. Part of this increase in heavy fuel prices reflects a change from high-sulfur to low-sulfur heavy fuel. The increases in coal and heavy fuel prices have contributed to the sharp increase in the imputed wholesale price of electricity on the East Coast.

While increases in wholesale gas prices during the past four years have ranged between 4 cents per Mcf and 16 cents per Mcf in the twenty-four SMSAs included in this study, increases in most of the SMSAs amounted to 10 to 11 cents per Mcf. These increases in the wholesale gas price covered all areas of the nation.

Of the various major primary sources of energy, distillate fuels showed the smallest relative increase in wholesale prices through 1972. The increases were widespread, however, covering all SMSAs where the use of this fuel was significant.

The increases in the dealer tankwagon gasoline price in the latter part of the 1960's and the early 1970's have been more erratic than those of most other wholesale fuel prices. This instability in the rise has been associated with the periods of widespread depressed pricing at the retail level. During depressed pricing episodes, major brand suppliers have protected their dealers by giving discounts or allowances off the dealer tankwagon price.

Table 3-18. Imputed Wholesale Price of Electricity (Mills Per Kilowatt Hour)

Census Region *SMSA*	*1960*	*1969*	*1972*
New England			
Boston, Mass.	10.5	8.6	15.8
Middle Atlantic			
New York, N.Y.	12.9	12.9	18.4
South Atlantic			
Charlotte, N.C.	5.8	5.6	7.9
East North Central			
Chicago, Ill.	9.3	8.5	11.0
East South Central			
Nashville, Tenn.	4.1	4.6	7.2
West North Central			
Minneapolis-St. Paul, Minn.	8.6	8.7	9.8
West South Central			
Houston, Tex.	5.4	4.5	4.9
Mountain			
Denver, Colo.	8.7	8.8	8.2
Pacific			
San Fran., Calif.	8.1	8.0	8.8

Source: See Table 8-b in Appendix.

Table 3-19. Wholesale Prices of Primary Fuels

Census Region SMSA	1960 Gas[a] ¢/Mcf	1960 No. 2 Fuel Oil[b] ¢/Gal.	1960 No. 6 Fuel Oil[b] $/Bbl.	1960 Coal[c] $/Ton.	1969 Gas[a] ¢/Mcf	1969 No. 2 Fuel Oil[b] ¢/Gal.	1969 No. 6 Fuel Oil[b] $/Bbl.	1969 Coal[c] $/Ton.	1972 Gas[a] ¢/Mcf	1972 No. 2 Fuel Oil[b] ¢/Gal.	1972 No. 6 Fuel Oil[b] $/Bbl.	1972 Coal[c] $/Ton
New England												
Boston, Mass.	57.4	9.8	2.67	—	63.9	11.0	2.10	—	75.4	11.9	4.76	—
Middle Atlantic												
New York, N.Y.	42.1	10.0	2.57	7.10	42.4	10.8	2.38	7.09	53.7	11.7	4.88	14.80
South Atlantic												
Charlotte, N.C.	33.0	10.4	—	4.75	34.0	11.6	—	6.92	44.0	11.5	—	14.10
East North Central												
Chicago, Ill.	31.7	9.5	3.26	4.35	29.7	10.5	3.36	5.22	40.9	11.0	4.62	7.00
East South Central												
Nashville, Tenn.	40.1	—	—	4.75	33.0	—	—	6.92	44.5	—	—	14.10
West North Central												
Minneapolis-St. Paul, Minn.	45.0	10.3	3.42	4.61	36.5	10.9	3.78	4.28	45.3	10.8	4.31	n.a.
West South Central												
Houston, Tex.	22.3	—	—	—	26.9	—	—	—	35.4	—	—	—
Mountain												
Denver, Colo.	27.5	—	—	5.55	25.8	—	—	4.52	32.9	—	—	4.64
Pacific												
Seattle, Wash.	31.9	10.6	2.78	7.54	32.5	11.4	2.88	8.21	41.3	12.7	4.26	n.a.

n.a.—not available.

[a] City Gate Prices.

[b] Average of postings for refinery or terminal of Jan. and July of given year and January of succeeding year.

[c] Fob mine price

Source: See Tables 8-a, 8-c, 8-d and 8-f in Appendix.

Table 3-20. Wholesale Prices* to Service Station Dealers of Major Brand Regular Grade Gasoline (Cents Per Gallon)

Census Region	*1960*		*1966*		*1972*		*1973*
SMSA	*January*	*July*	*January*	*July*	*January*	*July*	*January*
New England							
Boston, Mass.	9.9	14.4	15.1	15.5	19.4	17.3	18.0
Middle Atlantic							
New York, N.Y.	16.2	16.2	16.2	16.2	20.0	20.1	20.1
South Atlantic							
Charlotte, N.C.	15.6	14.9	15.8	16.2	18.7	17.4	18.8
East North Central							
Chicago, Ill.	16.4	16.4	16.9	16.9	17.5	14.7	19.6
East South Central							
Nashville, Tenn.	14.9	14.9	15.7	16.1	17.8	17.8	17.8
West North Central							
Minneapolis-St. Paul, Minn.	15.3	16.8	16.4	16.9	19.6	18.2	18.2
West South Central							
Houston, Tex.	15.9	15.9	14.6	15.1	17.7	16.3	16.3
Mountain							
Denver, Colo.	12.3	17.6	13.7	16.1	17.5	18.2	16.8
Pacific							
Los Angeles, Calif.	17.8	17.8	12.9	15.9	16.9	15.6	17.9

*Excludes excise taxes.

Source: See Table 8-e in Appendix.

Chapter Four

Special Analyses

Chapter four is devoted to an examination of different facets of energy pricing, including: (1) a breakdown of retail energy prices by cost component, (2) a comparison of space heating and gasoline costs in urban and rural areas, (3) the relationship of price to volume, (4) volume discounts and promotional rates, (5) interruptible gas rates, (6) spot versus contract prices, and (7) incremental gas pricing. The purpose of this chapter is to illuminate some of the more important aspects of energy pricing that a student of this field is likely to encounter.

A. BREAKDOWN OF RETAIL ENERGY PRICES BY COST COMPONENT

To provide insight into energy price developments, the retail prices of the various forms of energy to different types of customers in selected SMSAs for the years 1960 and 1971 have been broken down into these major cost components: production, transportation, distribution, and taxes.[a] Because of the differing structures of the energy industries, the components presented are not strictly comparable, but comparable enough to be meaningful.

Definitions

For the *production component* of the retail prices, the following were used:

Gas—the average price paid to producers in the field or at natural gasoline plants by the pipeline supplying the distributor in each

[a]See tables in the 9 series in the Appendix. It will be noted that a breakdown was ommitted for some retail prices for a particular form of energy in some SMSAs for some types of users because of a lack of necessary data or, in a few instances, because of technical problems.

SMSA. (Source: the Form 2 Annual Reports by interstate pipelines to the FPC.) If more than one pipeline supplied a distributor, the average price paid by each pipeline was weighted by the per cent of the distributor's total purchases supplied by the respective pipeline. In a few SMSAs where the supplier to the pipeline was another pipeline, the supplier pipeline's price was used. These instances have been noted on Table 9–f in the Appendix.

Electricity—the imputed wholesale cost per Kwh based on the sum of production costs of the utility's generating plants and an estimate of fixed costs divided by the Kwh generated.[a]

Coal—the F.O.B. price at the mine-mouth obtained from coal producers and the U.S. Bureau of Mines.

Motor Gasoline and

#2 Fuel Oil—the Gulf Coast cargo price for SMSAs east of the Mississippi River; the posted price for Oklahoma Group 3 northern shipments for SMSAs between the Rocky Mountain and the Mississippi River; and for Los Angeles and Seattle, the terminal prices in those areas discounted by the same relative difference existing between the Gulf Coast cargo price and the terminal price in Houston or New Orleans.[b] (Source of the price data was *Platt's Oilgram Price Service.*)

#6 Fuel Oil—the cargo price at East Coast points, where available, and for other SMSAs the posted price at the particular SMSA discounted by the average spread between the cargo price and the termi-

[a] See pages 68 and 69 on Electric Wholesale Price.

[b] The production component of light oil product price is the costs of the oil refineries, including the cost of their raw material, crude oil. While there are many refineries located in or very near to the SMSAs for which retail oil prices have been given in this study, the Gulf Coast cargo price and the Oklahoma Group 3 northern shipment posting were used for all SMSAs east of the Rockies for a variety of reasons, including: (1) the Southwest has been the major source of crude oil for this country and has furnished varying proportions of the crude oil needed by refineries in the North and East. Although in more recent years imported crude oil has supplied a large part of the East Coast refineries' needs, the Gulf Coast refineries have continued to furnish a substantial portion of the gasoline and distillate fuels used on the East Coast as well as a large amount of these products consumed in the Mid-west. (2) Some of the northern refineries use some crude oil production located much closer to these refineries than the Southwest crude oil production, which supplies refineries on the Gulf Coast and those posting prices for Oklahoma Group 3 northern shipments. Nevertheless, any difference in the transportation cost to these northern refineries for more proximate crude oil, assuming other qualities of the crude are the same, has tended to be offset by a higher price for these more accessible crude oils. (3) The terminal postings of the northern refineries located near SMSAs in this study are generally for truck transport or tank car lots, which are much smaller than the Gulf Coast cargo quantities or the pipeline shipments north from Oklahoma Group 3. While prices for these latter are not strictly refinery costs, they are much closer to a true refinery cost than the truck transport terminal postings.

nal price at selected East Coast points and the Gulf Coast.[a] (Price data were from *Platt's Oilgram Price Service.*)

The distinct functions of transportation and distribution are impossible to separate completely. Consequently, the definitions used for purposes of this study are somewhat arbitrary.

The *transportation components* of the retail prices of the different sources of energy were defined as follows in this study:

Gas—the difference between the average price the local distributor in a SMSA pays the pipeline suppliers for its gas and the average price the pipelines pay the producers in the field or natural gasoline plant (i.e., the difference between the city gate price) (for source, see note Table 8-A in Appendix) (and the production cost as defined above).

Electricity—transportation costs included with the distribution cost.

Coal—the transportation cost by rail or water shipment, whichever is lower, from the mine of the producer to the customer (Sources of data were the coal producers and users.)

Motor Gasoline

#2 Fuel Oil—for SMSAs east of the Rocky Mountains the cost of moving the products from the southwest refineries to the metropolitan centers, based on the pipeline tariff rates published by the Interstate Commerce Commission or tanker rates from *Platt's Oilgram Price Service.*

#6 Fuel Oil—transportation costs included with the distribution cost.

The *distribution costs* were defined and developed as follows:

Gas—the difference between the price paid by the retail customers, excluding taxes (source: American Gas Association *Rate Service*), and the average price paid by the local gas utility distributor to the pipeline at the city gate as discussed above.

Electricity—the difference between the prices paid by the retail customers, excluding taxes (source: FPC *National Electric Rate Book*). and the imputed cost of generating electricity or the calculated cost to the bus bar of the electric utility serving a SMSA as discussed above.

[a]Most of the residual fuel oil used on the East Coast is imported by tanker and that used in interior markets is generally produced by nearby refineries, except for some barge movements where water transportation is available. Overland transportation costs for heavy fuel oil are relatively high; residual fuel has not been susceptible to pipeline transportation over any substantial distance. As was noted for gasoline and distillate fuel oil, residual fuel oil truck transport or tank car refinery or terminal postings are much less indicative of the refinery cost than are cargo prices. Hence, cargo prices for East Coast points were used for production costs, and for other SMSAs in the study a discount was applied to the refinery or terminal transport or tank car postings in developing an estimated refinery cost.

Coal—no distribution costs by definition.

Motor Gasoline and

#2 Fuel Oil—the difference between the price to the ultimate customer, excluding taxes (for source see page 67), and the sum of the production and transportation costs as discussed previously.

#6 Fuel Oil—the difference between the price paid by the ultimate customer, excluding taxes (for source see page 67), and the cargo price at the pertinent East Coast SMSA or the imputed refinery cost for the appropriate inland points as discussed above.

The *tax component* of the price or cost to the ultimate consumer included any excise or sales taxes paid by the consumer for purchase or use of any of the forms of energy included in the study. Sources of tax information included the Commerce Clearing House, the American Petroleum Institute, the American Gas Association *Rate Service*, and the Federal Power Commission's *National Electric Rate Book*.

Developments

Production costs for all major sources of energy increased between 1960 and 1971 in most of the twenty-four SMSAs included in the study. On the other hand, transportation and distribution costs showed a varied pattern of change, particularly for gas and electricity, among the different types of users with increases in some SMSAs largely balanced by decreases in others. For oil products and coal, however, transportation and distribution costs generally rose. Excise and retail taxes were higher in most SMSAs, with hardly any SMSAs showing a reduction in taxes on energy.

As the data tabulated in Appendix Tables 9-a-1 through 9-f-4 demonstrate, the relative changes in the cost components of the retail price to the various types of users during the period 1960 to 1971 differed from SMSA to SMSA and among the types of energy. While production costs for a specific form of energy were the same for all customers in a particular SMSA, the transportation and distribution costs—and the taxes—varied with the form of energy and the type of user.

A breakdown of the prices of electricity for non-substitutable baseload use in homes (lighting, refrigerators, and small appliances) reveals that distribution costs represent the major cost component in all the SMSAs, but these costs became a relatively smaller proportion of the total between 1960 and 1971 (see Table 4-1). Production costs and taxes represented a larger share of the total retail price.

Table 4-1. Percent Breakdown of Energy Prices of Electricity Non-Substitutable Baseload by Cost Component

	1960	1971
New York City, N.Y.		
Production	31.5	42.3
Transportation & Distribution	66.6	49.9
Taxes	1.9	7.8
Total	100.0	100.0
Chicago, Illinois		
Production	30.6	32.7
Transportation & Distribution	64.1	59.8
Taxes	5.3	7.5
Total	100.0	100.0
Charlotte, N.C.		
Production	23.1	30.6
Transportation & Distribution	74.0	66.5
Taxes	2.9	2.9
Total	100.0	100.0
Los Angeles, Cal.		
Production	31.7	34.3
Transportation & Distribution	65.4	59.1
Taxes	2.9	6.6
Total	100.0	100.0

For residential heating, the relative importance of the various cost components differs with the type of fuel used (see Table 4-2). Around 55 per cent of the cost for fuel oil (#2 fuel oil) in all cities in which this fuel is important was accounted for by the production cost in 1960, but the per cent tends to be lower in 1971. Transportation and distribution costs of fuel oil, however, showed a substantially larger absolute gain than production costs and increased their share of the total retail price from 1960 to 1971. In contrast to fuel oil, the production cost of gas represents in most SMSAs a relatively small part (less than 20 per cent) of the total retail price for home heat, with transportation and distribution costs in 1971 accounting for around three-fourths of the total price. However, for gas home heat, transportation and distribution costs in 1971 represented a smaller per cent of the total price than in 1960. Electric heat prices, which frequently have been influenced by promotional rates, registered dramatic decreases in the distribution cost component in many SMSAs during the past decade. Production costs and the tax component in electric heat prices, on the other hand, rose both absolutely and relatively in most SMSA.

The breakdown of the cost components of retail energy prices to small commercial users (see Table 4–3) is similar to that of residential consumer prices. For electricity prices, however, the pattern tends to follow that for non-substitutable electric residential baseload rather than that for residential electric heating, which is heavily influenced by promotional rates.

Table 4–2. Percent Breakdown of Energy Prices for Home Heating[a] by Cost Component

	Electricity		*Gas*		*No. 2 Fuel Oil*	
	1960	*1971*	*1960*	*1971*	*1960*	*1971*
New York City, N.Y.						
Production	—	—	10.2	13.1	56.1	49.4
Transportation & Distribution	—	—	86.9	80.3	41.0	44.7
Taxes	—	—	2.9	6.6	2.9	5.9
Total	—	—	100.0	100.0	100.0	100.0
Chicago, Ill.						
Production	49.3	69.8	13.6	15.2	57.2	54.6
Transportation & Distribution	45.4	16.4	81.1	77.5	39.8	40.4
Taxes	5.3	13.8	5.3	7.3	3.0	5.0
Total	100.0	100.0	100.0	100.0	100.0	100.0
Charlotte, N.C.						
Production	37.6	64.7	12.1	18.0	57.1	57.9
Transportation & Distribution	59.5	32.4	85.0	79.1	40.0	38.2
Taxes	2.9	2.9	2.9	2.9	2.9	3.9
Total	100.0	100.0	100.0	100.0	100.0	100.0
Los Angeles, Cal.						
Production	53.0	63.3	13.2	15.7	—	—
Transportation & Distribution	44.1	27.6	83.9	76.0	—	—
Taxes	2.9	9.1	2.9	8.3	—	—
Total	100.0	100.0	100.0	100.0	—	—

[a] Electricity and Gas include substitutable base load.

Table 4-3. Percent Breakdown of Energy Prices to Small Commercial Users by Cost Component

	Electricity		*Gas*		*No. 2 Fuel Oil*	
	1960	*1971*	*1960*	*1971*	*1960*	*1971*
New York City, N.Y.						
Production	37.4	34.9	9.4	11.1	58.1	50.7
Transportation & Distribution	59.6	57.1	87.6	82.2	38.9	43.3
Taxes	3.0	8.0	3.0	6.7	3.0	6.0
Total	100.0	100.0	100.0	100.0	100.0	100.0
Chicago, Ill.						
Production	23.8	26.4	14.0	17.9	59.2	56.2
Transportation & Distribution	70.9	66.2	80.6	74.7	37.7	38.7
Taxes	5.3	7.4	5.4	7.4	3.1	5.1
Total	100.0	100.0	100.0	100.0	100.0	100.0
Charlotte, N.C.						
Production	23.5	30.3	11.8	16.0	59.1	59.7
Transportation & Distribution	73.6	66.8	85.2	81.0	37.9	36.3
Taxes	2.9	2.9	3.0	3.0	3.0	4.0
Total	100.0	100.0	100.0	100.0	100.0	100.0
Los Angeles, Cal.						
Production	24.6	28.0	16.1	18.9	—	—
Transportation & Distribution	72.5	65.4	81.0	72.8	—	—
Taxes	2.9	6.6	2.9	8.3	—	—
Total	100.0	100.0	100.0	100.0	—	—

Table 4-4. Percent Breakdown of Energy Prices to Large Commercial Users by Cost Component

	Electricity		*Gas*		*No. 2 Fuel Oil*		*No. 6 Fuel Oil*	
	1960	*1971*	*1960*	*1971*	*1960*	*1971*	*1960*	*1971*
New York City, N.Y.								
Production	66.4	51.1	11.5	14.3	79.2	74.9	75.8	71.6
Transportation and Distribution	30.7	41.1	85.6	79.1	18.0	19.4	21.2	22.7
Taxes	2.9	7.8	2.9	6.6	2.8	5.7	3.0	5.7
Total	100.0	100.0	100.0	100.0	100.0	100.0	100.0	100.0
Chicago, Ill.								
Production	59.1	55.1	21.7	23.4	82.7	79.4	80.5	75.4
Transportation and Distribution	35.6	37.0	72.9	69.3	14.3	15.8	16.7	19.8
Taxes	5.3	7.4	5.4	7.3	3.0	4.8	2.8	4.8
Total	100.0	100.0	100.0	100.0	100.0	100.0	100.0	100.0
Charlotte, N.C.								
Production	52.1	65.3	12.7	28.2	75.9	76.6	—	—
Transportation and Distribution	45.0	31.8	84.4	68.9	21.4	19.5	—	—
Taxes	2.9	2.9	2.9	2.9	2.7	3.9	—	—
Total	100.0	100.0	100.0	100.0	100.0	100.0	100.0	100.0
Los Angeles, Calif.								
Production	61.1	68.6	21.0	22.7	—	—	—	—
Transportation and Distribution	36.0	24.8	76.1	69.1	—	—	—	—
Taxes	2.9	6.6	2.9	8.2	—	—	—	—
Total	100.0	100.0	100.0	100.0	100.0			

For large commercial establishments, the production costs represent a much larger portion of the total energy price than for small commercial establishments; and conversely, the transportation and distribution costs are a much smaller portion. This situation prevails for all three types of energy used—electricity, gas, and fuel oil. The largest difference is in electricity, in which production costs in 1971 accounted for over half of the total price of electricity to the large commercial establishments, and only a little more than one-fourth of the total price to small commercial establishments in the SMSAs studied. The smallest differential in the relative importance of production costs between small and large commercial establishments is in gas prices.

As for small commercial establishments, the production component in energy costs to the large commercial establishment increased from 1960 to 1971 for both electricity and gas, with decreases ocurring in the relative importance of the transportation and distribution costs (see Table 4-4). Taxes became more important, although they remained a small part of the total. On the other hand, in fuel oil prices the relative importance of both production and transportation and distribution costs showed only small changes during the two years compared.

Of the three energy sources for large commercial establishments, the production costs' share of the total retail price is the highest for fuel oils and the lowest for gas. Conversely, the portion accounted for by transportation and distribution costs is the lowest for fuel oils and the highest for gas. Electricity is in-between fuel oil and gas. While there exists variation among the SMSAs in the relative importance of the cost components of retail prices to large commercial users, the pattern of relatively high production costs and relatively low transportation and distribution costs for oil and the opposite pattern for gas prevails for almost all the SMSAs studied. The production component of electricity prices to large commercial establishments generally is higher than the distribution component, but in most SMSAs the spread between these cost components is less than for fuel oil and gas, and in a number of the cities studied, the distribution component in 1971 exceeded the production cost.

The patterns evident in the relative importance and change in the cost components of energy prices to large commercial establishments are similar for large industrial establishments[a] For electricity prices, however, the production costs represented a substantially higher share of the total price and distribution an appreciably lower share than for large commercial establishments (see Table 4-5).

[a]In the breakdown of industrial energy prices by cost component, the price of firm gas was used. Undoubtedly the production component in industrial gas prices would be a higher share of the total retail price and the transportation and distribution costs a lower share if interruptible gas prices had been used.

Table 4-5. Percent Breakdown of Energy Prices to Large Industrial Users by Cost Component

	Electricity		*Gas*		*No. 2 Fuel Oil*		*No. 6 Fuel Oil*		*Coal*	
	1960	*1971*	*1960*	*1971*	*1960*	*1971*	*1960*	*1971*	*1960*	*1971*
New York City, N.Y.										
Production	88.7	72.2	12.4	15.4	81.4	78.1	78.1	75.9	58.1	67.5
Transportation and Distribution	10.3	20.0	87.6	84.6	18.6	21.9	21.9	24.1	41.9	32.5
Taxes	1.0	7.8	—	—	—	—	—	—	—	—
Total	100.0	100.0	100.0	100.0	100.0	100.0	100.0	100.0	100.0	100.0
Chicago, Ill.										
Production	94.2	82.2	24.1	25.3	82.7	79.4	80.5	75.3	56.3	57.9
Transportation and Distribution	0.5	10.4	70.6	67.3	14.3	15.8	16.7	19.8	41.8	37.6
Taxes	5.3	7.4	5.3	7.4	3.0	4.8	2.8	4.9	1.9	4.5
Total	100.0	100.0	100.0	100.0	100.0	100.0	100.0	100.0	100.0	100.0
Charlotte, N.C.										
Production	—	—	13.0	28.7	73.9	77.8	—	—	52.2	68.9
Transportation and Distribution	—	—	86.0	70.3	25.2	19.8	—	—	46.8	30.1
Taxes	—	—	1.0	1.0	0.9	2.4	—	—	1.0	1.0
Total	—	—	100.0	100.0	100.0	100.0	—	—	100.0	100.0
Los Angeles, Calif.										
Production	89.2	88.8	22.2	25.4	—	—	—	—	—	—
Transportation and Distribution	10.8	11.2	77.8	74.6	—	—	—	—	—	—
Taxes	—	—	—	—	—	—	—	—	—	—
Total[a]	100.0	100.0	100.0	100.0	—	—	—	—	—	—

[a]For electricity, the figures are those of the publicly-owned utility.

Since coal is a significant source of energy in large industrial plants in some areas, a breakdown of coal prices by cost component has been included for the industrial sector. Transportation and distribution costs of coal, while relatively lower than for gas, are considerably higher than for fuel oil. The relative importance of the production costs in coal prices increased in most SMSAs in the eastern half of the country between 1960 and 1971, a pattern similar to that of gas and electricity in most SMSAs of the nation.

Table 4-6. Percent Breakdown of Energy Prices of Gasoline at Service Stations by Cost Component

	1960	*1971*
New York City, N.Y.		
Production	34.5	28.3
Transportation & Distribution	32.0	39.4
Taxes	33.5	32.3
Total	100.0	100.0
Chicago, Ill.		
Production	33.4	29.3
Transportation & Distribution	36.8	36.8
Taxes	29.8	33.9
Total	100.0	100.0
Charlotte, N.C.		
Production	35.6	32.4
Transportation — Distribution	28.8	36.1
Taxes	35.6	31.5
Total	100.0	100.0
Los Angeles, Cal.		
Production	35.1	31.2
Transportation & Distribution	31.4	39.0
Taxes	33.5	29.8
Total	100.0	100.0

Retail gasoline prices present a unique pattern with respect to the various cost components, because the retail price is divided into three roughly equal parts—production cost, transportation and distribution costs, and taxes (see Table 4-6). Moreover, the variation among the nation's SMSAs is surprisingly small. The retail service station gasoline market is the only market in which taxes constitute a major portion of the total price.

While the absolute level of each of the major costs comprising retail gasoline prices rose between 1960 and 1971, the changes were not uniform. Consequently, changes in the relative importance of the components occurred. The relative importance of production costs and taxes generally decreased to a moderate extent in the retail gasoline price structure. On the other hand, distribution and transportation costs became a noticeably larger portion of the total price during this period.

B. COMPARISON OF SPACE HEATING AND GASOLINE COSTS IN URBAN AND RURAL AREAS

To investigate energy costs in rural versus urban areas, communities in twenty-four rural areas contiguous to the SMSAs were selected for a comparison of residential space heating costs for gas, electricity, and #2 fuel oil, as well as for gasoline. Moreover, in order to simplify the comparison, customers were assumed to use the same fuel for both the substitutable baseload and space heating.

The rural areas chosen are listed on the following page. Data sources were the AGA *Rate Service* for gas prices, the FPC *National Electric Rate Book* for electric prices, and the BLS *Retail Prices and Indexes of Fuel and Electricity, Fuel Oil and Oil Heat* and oil company information for urban fuel oil costs. Special material from various oil companies furnished the basis for the calculation of fuel oil prices in rural areas. Gasoline prices were obtained from *Platt's Oilgram Price Service* for most SMSAs; data for the remaining SMSAs and for the rural areas were furnished by oil companies. Space heating monthly bills are presented in Table 4-7 and gasoline prices in Table 4-8.

Space Heating Price Trends

While the results are not conclusive, it is still possible to perceive some general trends. For both gas and electricity, there has been a tendency for the rural-urban differentials to narrow slightly between 1960 and 1972 against a background of generally rising gas prices and declining electric prices.

Towns in Rural Areas Contiguous to Metropolitan Areas
(Metropolitan Areas in Parenthesis)

1. Plymouth, Massachusetts (Boston)
2. Voorhesville, New Yrok (Albany)
3. Waynesburg, Pennsylvania (Pittsburgh)
4. Middletown, New York (New York City)
5. Frederick, Maryland (Washington, D.C.)
6. Winder, Georgia (Atlanta)
7. Concord, North Carolina (Charlotte)
8. Stuart, Florida (Miami)
9. Franklin, Tennessee (Nashville)
10. Morris, Illinois (Chicago)
11. Howell, Michigan (Detroit)
12. Circleville, Ohio (Columbus)
13. Bay City, Texas (Houston)
14. Norco, Louisiana (New Orleans)
15. Claremore, Oklahoma (Tulsa)
16. Farmington, Missouri (St. Louis)
17. Shakopee, Minnesota (Minneapolis-St. Paul)
18. Blair, Nebraska (Omaha)
19. Fort Morgan, Colorado (Denver)
20. Globe, Arizona (Phoenix)
21. Logan, Salt Lake City)
22. El Centro, California (Los Angeles)
23. Merced, California (San Francisco)
24. Snohomish, Washington (Seattle)

In 1960 gas rates were available for twenty of the twenty-four rural areas.[a] The variation[b] between urban and rural bills was not large. In 65 per cent of the cases it was between 0 and 15 per cent. Larger variations existed for New York, Pittsburgh, Los Angeles, Seattle, Houston, and Omaha. In each of the latter two cities, however, one of the distributors was government-owned (the rural for Houston and the urban for Omaha), and this could account for some if not all of their larger variations.

[a] Those for which 1960 data were not available were the rural areas contiguous to Nashville, Denver, Miami, and Salt Lake City.

[b] Measured by the per cent by which urban monthly bills are above or below rural bills.

Table 4-7. Comparison of Typical Monthly Space Heating Bills for Residential Customers in Urban and Rural Areas (In Dollars)

	Gas				*Electric*				*Fuel Oil*	
	Urban		*Rural*		*Urban*		*Rural*		*Urban*	*Rural*
	1960	*1972*	*1960*	*1972*	*1960*	*1972*	*1960*	*1972*	*1972*	
New England										
Boston, Mass.	33.02	38.44	29.10	29.00	147.01	94.60	162.41	95.61[1]		n.a.
Middle Atlantic										
Albany, N.Y.	29.52	33.59	same		89.69	102.06	same		46.07	46.07
New York, N.Y.	27.16	29.28	20.56	26.14	94.97	77.88	66.90[2]	73.30		n.a.
Pittsburgh, Pa.	16.32	25.51	20.17	26.17[1]	83.19	64.85	93.21	67.34		n.a.
South Atlantic										
Atlanta, Ga.	14.34	17.53	15.55	16.41	46.50	36.58	same			n.a.
Charlotte, N.C.	19.49	16.83	20.17	18.43	58.42	41.51	45.37	46.73		n.a.
Washington, D.C.	29.88	31.77	26.60	26.86	84.75	58.79	112.46	51.69	31.32	30.26
Miami, Fla.	6.74	1.76[1]	n.a.	n.a.	9.55	9.02	same			n.a.
East South Central										
Nashville, Tenn.	20.96	24.00	n.a.	19.45	35.78	44.78	36.51	43.32		n.a.
East North Central										
Chicago, Ill.	24.41	28.28	21.78	22.43	93.68	72.22	89.46	68.87[1]	32.06	32.06
Detroit, Mich.	20.86	21.59	21.44	26.83	115.46	87.44	same		36.32	36.32
Columbus, Ohio	19.00	22.31	18.05	21.23	84.90	68.61	same			n.a.

West South Central											
Houston, Tex.	6.60	9.11	4.60	6.91		20.98	20.97	30.33	23.70		n.a.
New Orleans, La.	5.45	7.89	6.19	7.36		31.04	27.43	31.84	23.40		n.a.
Tulsa, Okla.	11.94	14.04	same			41.70	45.88	83.40	42.52		n.a.
West North Central											
Minneapolis-St. Paul, Minn.	24.48	26.26	22.15	26.02		99.56	84.79	98.09	72.16	34.99	34.99
Omaha, Neb.	16.47	17.18	23.03	24.43	govt.	81.43	50.57	110.75	56.46		n.a.
					private	94.96	76.32				
St. Louis, Mo.	13.50	17.66	16.06	19.84		52.06	31.51	77.13	42.84	25.29	24.65
Mountain											
Denver, Colo.	6.78	11.42	n.a.	10.42		100.90	62.91	46.25	50.94		n.a.
Phoenix, Ariz.	5.23	6.76	4.78	6.76		29.80	17.88	24.92	27.02[3]		n.a.
Salt Lake City, Utah	11.74	13.56	n.a.	13.56		91.49	66.20	88.55	82.17		n.a.
Pacific											
Los Angeles, Calif.	8.63	8.31	10.79	9.32	govt.	28.22	26.67	22.19	22.62		n.a.
					private	31.06	29.54				
San Francisco, Calif.	11.20	13.46	12.54	13.79		50.72	55.60	50.72	52.51		n.a.
Seattle, Wash.	30.30	24.92	20.83	24.92[1]	govt.	38.85	40.40	46.18	33.17		n.a.
					private	46.18	52.57				

Figures for gas and electric are for January 1 of the given year; for fuel oil are annual averages.

[1]Different utility.

[2]With gas substitutable base load would be $133.80.

[3]No winter rate as for Phoenix.

By 1972, while there was a slight increase in the number of metropolitan areas where prices exceeded nearby rural areas, the proportion with variations in the 0–15 per cent range rose so that the overall variation decreased slightly. Large variations were evident in Boston, Washington D. C., and Chicago (in addition to Houston and Omaha mentioned above and Nashville, where the rural distributor was government-owned).

Both rural and urban gas prices trend upward when the 1972 level is compared with that in 1960. The declines were few and small, while the increases usually exceeded 15 per cent. This upward thrust, which was stronger for rural communities, narrowed the proportionate spread between the two.

Electric rates are available for all areas for both periods of time. In 1960, for about one-fifth of the cities both urban and rural bills were the same. In the remainder, metropolitan area prices were usually lower, and the variations—whether positive or negative were substantial; over half exceeded 15 per cent.

There is no clearly defined regional pattern with regard to either size or direction except for the lack of much variation in the East Central region. It may be noted that urban costs were lower in the West Central region in spite of the prevalence of publicly owned rural distributors.

By 1972, urban electricity prices were higher than rural prices in more instances, not because they had risen more, but because they fell less. The variation was reduced, with 60 per cent being within the 0–15 per cent range. In most of the East and East Central areas the variation was small; in the West North Central and West South Central, while it was not small, it was less than in the earlier period. The Western cities registered the greatest variation and no clear pattern of change.

Electric bills trended downward in all cities but Albany, Nashville, Tulsa, San Francisco, and Seattle (in Houston there was no change). During this period, costs not only declined, but declined substantially, while whatever increases occurred were usually under 15 per cent. Rural prices exhibited a pattern of greater decline, thus reducing the difference between the two.

Fuel oil is significant as a source of energy for residential space heating in only ten of twenty-four SMSAs. Of these ten, data are available for only six rural areas in 1972. The sample tabulated indicates little rural-urban variance.

For retail gasoline prices, data for both rural communities and metropolitan centers could be obtained only for the recent period. Moreover, depressed pricing in the metropolitan centers tended to obscure whatever normal pattern may exist. If allowance is made for depressed pricing, however, it would appear that, based on the six city sample, gasoline prices in communities in rural areas near major metropolitan centers generally have the same retail gasoline prices as the urban centers. This phenomenon would be in harmony with the area pricing policies that many oil companies use.

Table 4-8. Comparison of Retail Service Station Gasoline Prices (Regular Grade) for Urban Centers and Towns in Rural Areas (cents per gallon)

SMSA	*Jan. 1972*		*Jan. 1973*	
	Urban	*Rural*	*Urban*	*Rural*
Charlotte, N. Car.	38.9	38.9	38.9	38.9
Washington, D.C.	37.9	36.9	38.9	38.9
Chicago, Ill.	37.3	40.4	40.4	40.4
Columbus, Ohio	37.9	37.9	37.9	37.9
Nashville, Tenn.	35.9	37.9	35.9	37.9
Minneapolis, Minn.	38.9	38.9	36.8	38.9
St. Louis, Mo.	36.9	35.9	38.9	37.9
Phoenix, Ariz.	37.9	38.9	36.9	38.9
San Francisco, Cal.	35.9	38.7	38.9	38.7
Seattle, Wash.	36.9	38.9	36.9	38.9

C. THE RELATIONSHIP OF PRICE TO VOLUME

In examining the relationship of energy prices to the volume purchased, the Chicago market has been used as an example because of the availability of data. Nevertheless, evidence exists to indicate that the patterns in the Chicago market are similar to those in other metropolitan areas of the nation.

The prices of the various forms of energy tend to decrease as the quantity purchased increases.[a] This phenomenon is apparent in the

[a] This pattern usually prevails for sales to the same type of customer. If comparisons are made between different classes of customers, however, the gas and electric schedules sometimes provide a lower rate for smaller quantities for residential use than for larger quantities for commercial use.

rate schedules of the gas and electric utilities supplying the Chicago market, in oil companies price postings[a] for #2 and #6 fuel oil, as well as in the typical prices reported by coal companies and retail dealers (see Table 4–9 on following page).

The price-volume relationship is not strictly linear, a break occurs between small quantities and large quantities. For instance, for natural gas the amount of the decline in price for firm volumes of 500 Mcf, 1000 Mcf, and 15,000 Mcf per month is approximately constant relative to the proportionate increase in the volume, but for a volume of 30 Mcf per month the price is significantly higher than would be indicated by an extrapolation of the relationship of the larger quantities. Nevertheless, the decline in the rate with the increase in volume tends to flatten out for very large volumes of gas. Electricity rates in Chicago record a pattern similar to that of gas. In the case of #2 fuel oil, the price is much higher per gallon for a partial truck load (the tankwagon price) than for a full truck transport (the transport price). A less pronounced variation prevails in the tankwagon and the transport price of #6 fuel oil. For coal, substantial breaks in the price-volume relationship occur between the small truck and the large rail carload deliveries, and between the rail carload and the unit train deliveries.

The difference in the cost per million Btu of small quantities of the primary fuels—gas, oil, and coal—is considerably greater than the difference in the cost of large quantities. The delivered prices of volumes of these fuels equivalent to 30 million Btu ranged in mid-December 1972 from 91.2 cents per MMBtu for #6 fuel oil (1 per cent sulfur maximum) to 130.5 cents per MMBtu for #2 fuel oil, with coal and gas within the range. On the other hand, for quantities equivalent to 2 billion Btu, the price in cents per MMBtu ranged from 74.4 cents for coal to 84.4 cents for #2 fuel oil.

D. VOLUME DISCOUNTS AND PROMOTIONAL RATES

The residential price tables in the Appendix present in most instances not just one monthly bill for each category of use, but an array. This diversity is the result of two factors—volume discounts and promotional

[a] Posted prices of fuel oil generally are not the same as actual prices, particularly for large volume contracts. In large volume contracts, however, prices historically have tended to vary from contract to contract, depending on a variety of factors such as the supply situation of the seller, the seller's appraisal of the competitive situation, the specific location of the buyer, the tankage of the buyer, the firmness and variability of the buyer's delivery requirements, and others. Posted prices were used in this study not only because of their availability but also because they provide at least a general indication of the price-volume relationship.

Table 4-9. Energy Costs per Volume Delivered Chicago, Illinois December 15, 1972

	Natural Gas Price[a]	
Volume	*Cents Per Mcf*	*Cents Per MMBtu*
30 Mcf per month	120	117
500 Mcf per month	95	92
1,000 Mcf per month	91	88
15,000 Mcf per month	78	76
45,000 Mcf per month	78	76
	# 2 Fuel Oil Price	
	Cents Per Gallon	*Cents Per MMBtu*
Tank Wagon: 1-399 gallons[b]	18.1	130.5
Tank Wagon: 400 gallons and over[b]	17.6	126.9
Transport: 6,000 gallons and over[c]	11.7	84.4
	# 6 Fuel Oil (1.00% Sulfur Maximum) Price	
	Cents Per Gallon	*Cents Per MMBtu*
Tank Wagon: 1-1,499 gallons	13.3	91.2
Tank Wagon: 1,500 gallons and over	12.8	87.7
Transport: 6,000 gallons and over	11.7	80.2
	Bituminous Coal Price	
	Dollars Per Ton	*Cents Per MMBtu*
Truck Delivery: 1 ton	26.98	103.5
Truck Delivery: 4-10 tons	26.20	100.8
Rail Delivery—1 car: 65-100 tons	19.34	74.4
Unit Train—100 cars: 9,000-10,000 tons	8.86	36.7
	Electricity Price	
	Cents Per Kwh	*Cents Per MMBtu*
2,000 Kwh per month	4.282	1,255
200,000 Kwh per month	2.122	622
2,000,000 Kwh per month	1.435	421
21,000,000 Kwh per month	1.409	413

[a] Firm contracts.

[b] Includes 0.5 cents per gallon for stamps.

[c] FOB refinery or terminal plus estimated delivery cost.

Source: Natural Gas—American Gas Association
2 Fuel Oil—*The Oil Daily,* December 18, 1972
6 Fuel Oil—*The Oil Daily,* December 18, 1972
Coal—Relevant coal companies and retail coal dealers
Electricity—Federal Power Commission*Electric Rate Book*

rates. Utility rates are usually structured so that a constant price is charged for all energy consumed in the first "block" and a lower price per unit is charged for all energy comsumed in succeeding blocks.

For example, the residential gas rates for Denver, Colorado, served by Public Service of Colorado and effective July 7, 1972, were as follows:

First 600 cu. ft. or less	$1.08
Next 1,600 cu. ft.	$0.10454 per hundred cu. ft.
Next 6,000 cu. ft.	$0.05654 per hundred cu. ft.
Next 6,000 cu. ft.	$0.04954 per hundred cu. ft.
Next 6,000 cu. ft.	$0.04754 per hundred cu. ft.
All additional	$0.04554 per hundred cu. ft.

Hence, a customer who has gas space heating but an electric substitutable baseload (water heating, range, and clothes dryer) would pay a higher heating bill, since it would be calculated from the beginning of the rate instead of after the 4,000 cu. ft. (4 Mcf) that would be used for the baseload.

In a few instances, the price for the last block is higher. This was the case in Nashville, which used TVA rates up to 1969, when they were discontinued. In Seattle the same pattern prevails currently, but it is atypical.

A customer can additionally qualify for different rates depending on the type of equipment in his home. An electric utility may charge less for all levels of use for an all-electric home in which electricity is the sole source of power for cooking, water heating, and space heating. Moreover, with electric air conditioning producing a summer peak in demand, many electric utilities offer promotional rates for heating use to help achieve a higher average load factor. The reverse load pattern characterizes gas consumption, where gas space heating is common and lower rates are offered for other uses to raise summer consumption.

Many kinds of promotional rates are available. Utilities have not been consistent in the types offered throughout the period under study. In general, simpler structures were employed in earlier years. The most popular with gas distributors has been the summer-winter rate with lower summer charges to encourage the use of gas for air conditioning. As a rule, winter heating prices are the same irrespective of whether a residence has gas air conditioning. An exception is Miami, where a year-round rate is offered, and air conditioning is a prerequisite in order to qualify for lower heating prices. An "all-gas" home rate is available in Seattle, where a customer must have gas for both space heating and water heating.

The most typical promotional rate structure for electric utilities is that offering lower prices to large volume users for such human need purposes as water heating and/or space heating. For water heating service some companies add on an end floating block either of fixed size or of an amount varying with tank size to the residential block rate. Others have special water heating and/or space heating rates separate from that for basic residential use covering the non-substitutable baseload.

The all-electric home design permits a separate and lower series of prices for levels of residential use, provided electricity is the sole source of power in the home. Like gas distributors, electric companies also offer a summer-winter rate, with the winter rate cheaper than the summer rate.

E. INTERRUPTIBLE GAS RATES FOR INDUSTRIAL MARKETS

In order to reduce costs, a significant proportion of which are fixed, natural gas distributors attempt to maximize the use of facilities while protecting firm loads. This is achieved by the employment of such devices as peak shaving, storage, and interruptible gas sales. Interruptible sales are those to industry with the provision that the service can be curtailed in order to accommodate the needs of firm customers, usually in the winter. Those industrial customers requiring gas for other than its Btu content, or for use in processing operations that must be maintained on a continuous basis, often pay a premium for firm purchases. Because of its intrinsic properties gas is the most suitable fuel for burning brick, for some operations in glass manufacture, and for heat-treating of metals, to name a few examples. The prices of gas to industrial customers are presented in Table 6–a in the Appendix. Firm gas rates exceed interruptible rates by widely varying margins.

Since interruptible users require alternate fuel burning facilities, the prices of these competitive fuels can be expected to influence the determination of intertuptible gas rates. In order to find buyers who will accept gas on an interruptible basis, the distributor has had to offer this gas at rates at least competitive with alternate fuels. During the earlier years of this study, coal or oil were either cheaper or virtually on a par with natural gas in two-thirds of the SMSAs.

The competitive situation changed by 1973, however, with gas prices clearly lower in most SMSAs. The spread between firm and interruptible rates, which was larger in the earlier years, narrowed significantly. In Tulsa and the California SMSAs, where alternate fuels provided little competition, the spread showed little change.

A comparison of firm and interruptible gas rates for January 1, 1970 is shown on Table 4–10 on the next page:

Table 4-10. Prices Paid by Industrial Customers for Firm and Interruptible Gas Purchases 1960 and 1973 (c/Mcf)

	1960			1973		
	Firm	*Inter-Ruptible*	*Differ-ential*[2]	*Firm*	*Inter-ruptible*	*Differ-ential*[2]
New England						
Boston	109.7	—	—	198.7	109.6	82
Middle Atlantic						
Albany	101.9	71.9	42	117.8	—	—
New York	127.3	36.0	254	148.6	52.2	185
Pittsburgh	51.2	—	—	77.7	—	—
South Atlantic						
Atlanta	62.9	24.4	158	90.9	43.0	109
Charlotte	121.1	42.0	188	81.0	49.0	65
Washington, D.C.	96.8	—	—	110.1	82.7	33
Miami	92.7	37.5	147	53.7	50.4	7
East South Central						
Nashville	52.8	30.1	75[3]	71.7	43.0	67
East North Central						
Chicago	64.0	33.3	92	83.4	49.3	69
Detroit	57.0	42.1	35	73.8	57.6	28
Columbus	56.8	—	—	74.7	—	—
West South Central						
Houston	19.0	—	—	n.a.	—	—
New Orleans	14.5	—	—	34.6	—	—
Tulsa	23.3	21.3	9	32.7	30.7	7
West North Central						
Minneapolis	68.2	—	—	76.7	—	—
Omaha	—	29.0	—	48.0	35.0	37
St. Louis	34.0	28.4[1]	20	62.1	47.0	32
Mountain						
Denver	29.5	9.6	207	46.3	21.7	113
Phoenix	45.0	—	—	58.4	—	—
Salt Lake City	29.8	22.3	34	33.4	27.9	20
Pacific						
Los Angeles	62.6	40.7	54	72.5	46.8	55
San Francisco	55.5	42.5	31	59.3	44.0	35
Seattle	67.7	39.7	71	85.7	54.5	57

[1]Price on July 1; January 1 not available.

[2]Per cent by which price of firm gas exceeds interruptible.

[3]Different utility in 1960 than in 1973. The differential in 1962, under the new utility, was 36.

Note: Where interruptible prices are not listed, they are not offered by the selected utility. Prices are those in effect on January 1 of the given year unless otherwise noted.

Source: See Table 6-a in Appendix.

F. SPOT VERSUS CONTRACT PRICES

Fuel Oil

Most fuel oils are marketed under contract—either formal contract or otherwise. Thus, based on the volume sold, the contract price is more representative of the price at which fuel oils are moving in the market than is the spot price. The spot price, however, is a much more sensitive indicator of market conditions than the contract price. Spot prices show larger gains than contract prices when the market is tightening, and they decline more when the market weakens.

Fuel oil spot prices generally are higher than contract prices, with the spread tending to widen when the market tightens. The tight supply situation in fuel oils in the winter of 1972–73 is reflected in the marked spread between spot and contract cargo prices in the New York Harbor shown on the following table:

Table 4-11. Estimated New York Harbor Cargo Prices[a] January 14, 1973

	Spot	*Net Contract*
#2 Heating Oil (cents per gallon)	13.25¢	10.90¢
#6 Fuel Oil (dollars per barrel)		
0.3% sulfur, low pour	$5.00	$4.40
1.0% sulfur, low pour	$3.95	$3.80

Source: *Platt's Oilgram Price Service*
[a]Duty Paid

During a weak market, however, spot prices sometimes fall below contract prices, particularly when storage limitations force a refiner to make distress sales.

Steam Coal

The majority of coal produced, like fuel oil, is sold on a contract basis. Many mines, however, will sell part of their output on a contract basis and the remainder on a spot basis in order to take advantage of any market opportunities. Spot purchases are often made by a user to fulfill a certain portion of his energy requirements or to meet unanticipated energy requirements created by increased production of products or services.

Spot steam coal prices reflect the market situation at a particular point in time. Since spot prices are heavily influenced by current market conditions, they are given to fluctuating widely above or below the contract price. Contract prices reflect market conditions at the time the contract was negotiated and therefore would remain throughout the life of the contract, or would be adjusted through escalation provisions in the contract. The movements of the contract price would not be given to any wide degree of fluctuation.

The following table sets out spot and contract prices for the industrial and power plant sectors for those SMSAs where data permitted such a comparison.

Table 4-12. Spot vs. Contract Prices for Steam Coal First Quarter, 1973 (In Dollars Per Ton)

	Industrial Sector		*Power Plant Sector*	
	Spot Price	*Contract Price*	*Spot Price*	*Contract Price*
St. Louis	$10.40	$ 9.52		
Chicago	$13.19	$11.30		
Minneapolis/ St. Paul			$10.56	$5.78
Denver			$ 6.75	$6.32
Atlanta			$ 9.93	$8.60
Detroit			$12.10	$9.57

Source: Foster Associates, and FPC Form 423, January, 1973

The spot price, in Table 4–12 above, is higher than the contract price in all areas. This results from factors also affecting other energy industries, such as increased costs and supply limitations, which would cause the current price of fuel to increase above the contract price.

G. INCREMENTAL GAS PRICING

In a decision issued June 28, 1972[a] the FPC authorized three U.S. interstate gas pipelines to import LNG from El Paso Algerian Corp. subject to several conditions, one of which was incremental pricing. Specifically, the FPC directed that the three importing pipelines sell the LNG to their customers at incremental rates instead of at rolled-in rates, as proposed.

[a] FPC Opinion No. 622, issued June 28, 1972 *in re Columbia LNG Corp. et al.* (CP71-68 et al.)

Incremental rates are based on the cost of the incremental source of supply alone— which in this case is higher than the cost of the pipelines' other supplies—while rolled-in rates reflect an averaging of the costs of all supplies. The Commission argued that rolled-in pricing of the more expensive LNG would disguise its true economic cost. Moreover, to assure that incremental LNG prices would be passed through to ultimate consumers, the FPC prohibited the three importing pipelines from selling to distributors that did not have separate rate schedules for reselling the LNG at incremental rates.

This condition was not only denounced by all the participants in the El Paso Algerian project, but it also evoked a considerable outcry from many other companies and interests.[a] With very few exceptions, pipelines and distributors agreed that incremental pricing of individual sources of supply would create overwhelming administrative and practical problems, particularly for distributors, and hence would be unworkable. State commissions argued that the FPC would be improperly regulating intrastate commerce.

Faced with such overwhelming opposition, the FPC modified its decision on October 5, 1972 (Opinion N. 622-A). Specifically, the Commission retained the requirement that the importing pipelines sell LNG to distributors on an incremental basis, but eliminated the further condition requiring distributors to sell also to retail customers on that basis.

[a]Because of the precedent-setting nature of Opinion No. 622, a large number of companies or groups not previously parties to the proceeding—including gas pipelines and gas distributors associations, several individual pipelines and distributors, Interior Department, Maritime Administration, AFL-CIO, National Association of Regulatory Utility Commissioners, and others—asked to intervene and filed petitions for rehearing.

Appendix Tables

Table 1-a

TYPICAL MONTHLY FUEL BILLS FOR RESIDENTIAL CUSTOMERS
BOSTON
ALL FUELS
($/Month)

	1960		1961		1962		1963		1964		1965	
	Jan.	July	Jan.	July	Jan.	July	Jan.	July	Jan.	July	Jan.	July
Non-Substitutable Base Load												
Electric												
with gas sub. base load								)14.49	14.49	14.49	13.81	13.81
with elec. sub. base load	14.29	14.44	14.44	14.60	14.60	14.54	14.54)					
without elec. space heat.								)11.98	11.98	11.98	11.36	11.36(
with elec. space heat.								)10.58	10.58	10.58	10.58	10.58(
Substitutable Base Load												
Electric												
without elec. space heat.								)16.05	16.05	16.05	16.80	16.80
with elec.	24.10	24.39	24.39	24.72	24.72	24.60	24.60)					
space heat.								)14.85	14.85	14.85	14.40	14.40
Gas	10.36	10.36	10.53	10.53	10.16	10.16	10.16	10.16	10.37	10.34	10.33	10.33
Space Heating												
Electric	147.01	148.80	148.80	150.78	150.78	150.09	150.09	90.61	90.61	90.61	87.82	87.82
Gas												
with elec. sub. base load	37.67	37.67	38.62	38.62	36.56	36.56	36.56	36.56	37.73	37.58	37.51	37.51
with gas sub. base load	33.02	33.02	33.97	33.97	31.91	31.91	31.91	31.91	33.09	32.93	32.86	32.86
Fuel Oil	25.59		26.76		26.26		26.91		26.64		27.48	

Note: Gas and electricity are based on prices in effect on January 1 and July 1 of the given year.

Source: American Gas Association Rate Book; FPC National Electric Rate Book.
Sources and procedures for oil estimates are described in Chapter IV.

1966		1967		1968		1969		1970		1971		1972		1973
Jan.	July	Jan.	July	Jan.	July	Jan.	July	Jan.	July	Jan.	July	Jan.	July	Jan.
13.17	13.17	13.40	13.40	13.40	13.40	13.40	13.40	13.40	13.40	13.40	15.02	14.75	14.85	14.50
10.73	10.73	10.96	10.96	10.96	10.96	10.96	10.96	10.96	10.96	10.96	12.58	12.31	12.41	11.95
15.15	15.15	15.60	15.60	15.60	15.60	15.60	15.60	15.60	15.60	15.60	18.84	18.31	18.51	17.78
12.53	12.53	12.98	12.98	12.98	12.98	12.98	12.98	12.98	12.98	12.98	16.22	15.69	15.89	15.15
10.33	10.33	10.28	10.30	10.30	10.30	10.30	10.84	10.84	11.51	11.51	13.03	13.03	13.03	14.09
75.31	75.31	78.08	78.08	78.08	78.08	78.08	78.08	78.08	78.08	78.08	97.86	94.60	95.81	91.36
37.51	37.51	37.51	37.24	37.04	37.04	37.04	38.86	38.86	39.53	39.53	45.11	45.11	45.11	49.58
32.86	32.86	32.60	31.81	31.81	31.81	31.81	33.09	33.09	33.08	33.08	38.44	38.44	38.44	41.72
28.61		29.30		29.74		30.29		32.00		33.75		34.56		

Table 1-b

TYPICAL MONTHLY FUEL BILLS FOR RESIDENTIAL CUSTOMERS
ALBANY
ALL FUELS
($/Month)

	1960		1961		1962		1963		1964		1965	
	Jan.	July	Jan.	July	Jan.	July	Jan.	July	Jan.	July	Jan.	July
Non-Substitutable Base Load												
Electric	9.32	9.32	9.32	9.32	9.52	9.52	9.61	9.61	9.61	9.86	9.95	9.95
Substitutable Base Load												
Electric	11.78	11.78	11.78	11.78	12.20	12.20	12.31	12.31	12.31	12.80	12.93	12.93
Gas	8.02	8.02	7.65	7.65	7.65	7.65	7.59	7.54	7.61	7.57	7.65	7.65
Space Heating												
Electric	89.69	89.69	89.69	89.69	92.82	92.82	93.72	93.72	93.72	97.25	98.19	98.19
Gas												
with elec. sub. base load	33.33	33.33	31.32	31.32	31.32	31.32	30.70	30.36	30.80	30.58	30.91	30.93
with gas sub. base load	29.52	29.52	27.42	27.42	27.42	27.42	26.76	26.42	26.86	26.63	26.93	26.95
Fuel Oil	29.82		31.02		31.77		32.82		32.82		33.62	

Note: Gas and electricity are based on prices in effect on January 1 and July 1 of the given year.

Source: American Gas Association <u>Rate Book</u>; FPC <u>National Electric Rate Book</u>. Sources and procedures for oil estimates are described in Chapter IV.

	1966		1967		1968		1969		1970		1971		1972		1973
	Jan.	July	Jan.	July	Jan.	July	Jan.	July	Jan.	July	Jan.	July	Jan.	July	Jan.
	9.95	9.95	10.05	10.05	10.05	9.85	9.85	9.89	10.60	10.60	10.70	10.70	10.76	12.51	12.40
	12.93	12.93	13.05	13.05	13.05	12.66	12.66	12.75	13.18	13.18	13.28	13.28	13.39	15.35	15.14
	7.65	7.65	7.73	7.73	7.74	7.73	9.09	9.08	9.21	9.34	9.49	9.65	9.82	10.00	10.13
	98.19	98.19	99.12	99.12	99.12	96.35	96.35	97.03	100.25	100.25	101.09	101.09	102.06	117.05	115.27
	30.92	30.93	31.19	31.19	31.27	31.21	34.68	34.62	35.53	36.38	37.51	38.57	39.79	41.21	41.93
	26.94	26.96	27.18	27.18	27.25	27.20	28.52	28.46	29.36	30.22	31.31	32.37	33.59	34.82	35.73
	34.48		36.12		38.49		39.08		41.20		45.34		46.07		

Table 1-c

TYPICAL MONTHLY FUEL BILLS FOR RESIDENTIAL CUSTOMERS
NEW YORK CITY
ALL FUELS
($/Month)

	1960		1961		1962		1963		1964		1965	
	Jan.	July	Jan.	July	Jan.	July	Jan.	July	Jan.	July	Jan.	July
Non-Substitutable Base Load												
Electric												
with gas sub. base load / with elec. sub. base load	16.08	15.79	15.79	15.79	15.93	16.01	16.17	15.99	15.99	15.99	16.96	16.97
Substitutable Base Load												
Electric												
without elec. space heat. / with elec. space heat.	17.54	17.14	17.14	17.14	17.48	17.54	17.72					
without elec. space heat.								17.38	17.38	17.38	18.44	18.45
with elec. space heat.								14.38	14.38	14.38	14.67	14.68
Gas	9.41	9.41	8.86	8.94	8.99	8.94	9.04	8.83	8.84	8.96	9.05	9.03
Space Heating												
Electric	94.97	92.75	92.75	92.75	92.76	92.45	95.22	65.30	65.30	65.30	66.98	67.04
Gas												
with elec. sub. base load	31.14	31.14	28.48	28.81	29.09	28.83	29.16	28.20	28.23	28.92	29.17	29.10
with gas sub. base load	27.16	27.16	24.54	24.87	25.77	24.90	25.19	24.23	24.25	24.96	25.17	25.11
Fuel Oil	22.63		23.44		23.41		23.96		23.42		23.83	

1/ All electric home rate 1971 to present.

Note: Gas and electricity are based on prices in effect on January 1 and July 1 of the given year.

Source: American Gas Association Rate Book; FPC National Electric Rate Book. Sources and procedures for oil estimates are described in Chapter IV.

1966		1967		1968		1969		1970		1971		1972		1973
Jan.	July	Jan.	July	Jan.	July	Jan.	July	Jan.	July	Jan.	July	Jan.	July	Jan.
16.97	16.91	20.80	20.88	20.88	21.08	21.08	21.15	21.15	21.15	)14.98	15.31	16.26	16.93	17.11
										)14.53[1/]	14.86	15.80	16.54	16.66
18.45	18.34	18.94	19.05	19.05	15.57	15.57	15.68	15.68	15.68	17.71	18.30	20.20	21.67	21.91
14.68	14.58	14.80	15.12	15.12	14.30	14.30	14.40	14.40	14.40	12.04	12.48	14.39	15.86	16.09
9.03	8.99	9.12	8.99	8.95	8.99	9.06	9.14	9.12	9.02	9.32	9.44	10.21	10.43	10.44
67.04	66.44	65.89	67.69	67.69	58.60	58.60	59.21	59.21	59.21	65.14	67.15	77.88	85.83	87.13
29.10	28.83	29.32	28.75	28.50	28.75	29.07	29.45	29.32	28.87	30.20	30.75	33.60	34.65	34.69
25.11	24.84	25.28	24.72	24.46	24.72	25.04	25.41	25.28	25.85	26.13	26.68	29.28	30.32	30.37
24.47		25.24		26.00		26.69		27.98		30.01		30.59		

Table 1-d
TYPICAL MONTHLY FUEL BILLS FOR RESIDENTIAL CUSTOMERS
PITTSBURGH
ALL FUELS
($/Month)

	1960		1961		1962		1963		1964		1965	
	Jan.	July	Jan.	July	Jan.	July	Jan.	July	Jan.	July	Jan.	July
Non-Substitutable Base Load												
Electric												
with gas sub. base load	10.09	10.09	10.09	10.09	10.09	10.09	10.09	10.09	10.09	10.09	9.68	9.68
with elec. sub. base load (without elec. space heat. / with elec. space heat.)	9.73	9.73	9.73	9.73	9.73	9.73	9.73	9.73	9.73			
without elec. space heat.										)8.26	8.26	8.26
with elec. space heat.										)8.11	8.11	8.11
Substitutable Base Load												
Electric												
without elec. space heating	16.22	16.22	16.22	16.22	16.22	16.22	16.22	16.22	16.22	13.18	13.18	13.18
with elec. space heating	13.08	13.08	13.08	13.08	13.08	13.08	13.08	13.08	13.08	10.55	10.55	10.55
Gas	4.92	4.92	4.92	4.91	4.91	5.15	5.13	5.13	5.12	5.18	5.18	5.17
Space Heating												
Electric												
with gas sub. base load / with elec. sub. base load	83.19	83.19	83.19	83.19	83.19	83.19	83.19	83.19	83.19	68.74	68.74	68.74
Gas												
with elec. sub. base load	18.59	18.59	18.59	18.40	18.40	19.45	19.37	19.32	19.30	19.69	19.68	19.66
with gas sub. base load	16.32	16.32	16.32	16.13	16.13	16.93	16.85	16.80	16.78	17.19	17.18	17.16
Fuel Oil	--	--	--	--	--	--	--	--	--	--	--	--

Note: Gas and electricity are based on prices in effect on January 1 and July 1 of the given year.

Source: American Gas Association Rate Book; FPC National Electric Rate Book.

1966		1967		1968		1969		1970		1971		1972		1973
Jan.	July	Jan.	July	Jan.	July	Jan.	July	Jan.	July	Jan.	July	Jan.	July	Jan.
9.68	9.68	9.68	9.68	9.68	9.68	9.74	9.74	9.74	10.77	10.77	11.33	11.88	11.96	11.94
8.26	8.26	8.26	8.26	8.26	8.26	8.32	8.32	8.32	9.19	9.19)	) 10.68	11.23	11.32	11.30
8.11	8.11	8.11	8.11	8.11	8.11	8.16	8.16	8.16	9.02	9.02)				
13.18	13.18	13.18	13.18	13.18	13.18	13.27	13.27	13.27	14.60	14.60	13.17	14.27	14.43	14.39
10.55	10.55	10.55	10.55	10.55	10.55	10.61	10.61	10.61	11.72	11.72	8.94	10.04	10.20	10.16
5.23	5.23	5.23	5.23	5.23	5.34	5.80	5.84	6.19	6.40	6.48	6.68	6.70	6.73	7.27
68.74	59.92	59.92	59.92	59.92	59.92	60.30	60.30	60.30	66.61	66.61)	)64.35	72.12	73.26	72.95
											)57.08	64.85	65.99	65.69
20.95	20.95	20.95	20.95	20.95	21.01	22.15	22.37	25.41	26.26	26.80	28.12	28.24	28.47	31.44
18.50	18.50	18.50	18.50	18.50	18.56	20.57	20.79	22.77	23.53	24.07	25.39	25.51	25.74	28.62
--	--	--	--	--	--	--	--	--	--	--	--	--	--	--

Table 1-e
TYPICAL MONTHLY FUEL BILLS FOR RESIDENTIAL CUSTOMERS
ATLANTA
ALL FUELS
($/Month)

	1960		1961		1962		1963		1964		1965	
	Jan.	July	Jan.	July	Jan.	July	Jan.	July	Jan.	July	Jan.	July
Non-Substitutable Base Load												
Electric												
without elec. space heat.	7.95	7.95	7.96	7.96	7.83	7.83	7.83	7.83	7.83	7.83	7.70	7.70
with elec. space heat.	6.72	6.72	6.73	6.73	6.59	6.59	6.59	6.59	6.59	6.59	6.18	6.18
Substitutable Base Load												
Electric												
without elec. space heat.	12.20	12.20	12.22	12.22	11.95	11.95	11.95	11.95	11.95	11.95	9.79	9.79
with elec. space heat.	11.58	11.58	11.60	11.60	11.33	11.33	11.33	11.33	11.33	11.33	10.97	10.97
Gas	4.36	4.36	4.58	4.58	4.58	4.58	4.58	4.85	4.85	4.85	4.85	4.85
Space Heating												
Electric												
with gas sub. base load with elec. sub. base load.2/	46.50	46.50	46.60	46.60	45.37	45.37	45.37	45.37	45.37	45.37	45.37	45.37
Gas												
with elec. sub. base load	15.48	15.48	15.48	15.48	15.48	15.48	15.48	15.48	15.48	15.48	15.48	15.48
with gas sub. base load	14.34	14.34	14.34	14.34	14.34	14.34	14.34	14.34	14.34	14.34	14.34	14.34
Fuel Oil	--	--	--	--	--	--	--	--	--	--	--	--

1/ All electric home rates no longer available.

2/ All electric home rates to 1972.

Note: Gas and electricity are based on prices in effect on January 1 and July 1 of the given year.

Source: American Gas Association Rate Book; FPC National Electric Rate Book.

1966		1967		1968		1969		1970		1971		1972		1973
Jan.	July	Jan.	July	Jan.	July	Jan.	July	Jan.	July	Jan.	July	Jan.	July	Jan.
7.70	7.70	7.70	7.70	7.70	7.70	7.87	7.87	7.87	7.87	8.03	8.03	9.11	9.15	9.57
6.18	6.18	6.18	6.18	6.18	6.18	6.31	6.31	6.31	6.31	6.48	6.48	1/		
9.79	9.79	9.79	9.79	9.79	9.79	10.00	10.00	10.00	10.00	10.33	10.33	11.45	11.54	12.08
10.97	10.97	10.97	10.97	10.97	10.97	11.21	11.21	11.21	11.21	11.55	11.55	1/		
4.85	4.85	4.85	4.85	4.99	4.99	4.99	4.99	4.99	5.57	5.67	6.32	6.39	6.34	6.36
45.37	45.37	45.37	45.37	45.37	45.37	46.36	46.36	46.36	46.36	47.91	47.91	)36.58	36.98	38.74
												)40.28	40.65	42.61
15.48	15.48	15.48	15.48	16.10	16.10	16.10	16.10	16.10	18.29	18.73	19.48	19.71	19.53	19.60
14.34	14.34	14.34	14.34	14.34	14.34	14.34	14.34	14.34	16.74	17.17	17.29	17.53	17.35	17.43
--	--	--	--	--	--	--	--	--	--	--	--	--	--	--

Table 1-f

TYPICAL MONTHLY FUEL BILLS FOR RESIDENTIAL CUSTOMERS
CHARLOTTE
ALL FUELS
($/Month)

	1960		1961		1962		1963		1964		1965	
	Jan.	July	Jan.	July	Jan.	July	Jan.	July	Jan.	July	Jan.	July
Non-Substitutable Base Load												
Electric												
with gas sub. base load	9.64	9.64	9.64	9.64	9.64	9.64	9.64	9.64	9.64	9.54	9.54	9.54
with elec. sub. base load												
without elec. space heat.	8.61	8.61	8.61	8.61	8.61	8.61	8.61	8.61	8.61	)8.23	8.23	8.23
with elec. space heat.										)8.51	8.51	8.51
Substitutable Base Load												
Electric												
without elec. space heat.	16.48	16.48	16.48	16.48	16.48	16.48	16.48	16.48	16.48	11.95	11.95	11.95
with elec. space heat.	12.36	12.36	12.36	12.36	12.36	12.36	12.36	12.36	12.36	11.43	11.43	11.43
Gas	8.01	8.01	8.17	8.17	8.17	8.17	7.89	7.89	7.89	7.89	7.68	7.68
Space Heating												
Electric												
with gas sub. base load	58.42	58.42	58.42	58.42	58.42	58.42	58.42	58.42	58.42	)54.67	54.67	54.67
with elec. sub. base load										)45.73	45.73	45.73
Gas												
with elec. sub. base load	22.97	22.97	23.67	23.67	23.67	23.67	22.37	22.37	22.37	22.37	20.60	20.60
with gas sub. base load	19.49	19.49	20.20	20.20	20.20	20.20	18.87	18.87	18.87	18.87	16.83	16.83
Fuel Oil	19.24		19.96		19.68		20.36		18.84		18.84	

Note: Gas and electricity are based on prices in effect on January 1 and July 1 of the given year.

Source: American Gas Association Rate Book; FPC National Electric Rate Book. Sources and procedures for oil estimates are described in Chapter IV.

1966		1967		1968		1969		1970		1971		1972		1973
Jan.	July	Jan.	July	Jan.	July	Jan.	July	Jan.	July	Jan.	July	Jan.	July	Jan.
9.54	9.54	9.54	9.54	9.54	9.54	9.54	9.54	9.54	9.80	9.80	9.80	9.80	11.46	11.46
8.23	8.23	8.23	8.23	8.23	8.23	8.23	8.23	8.23	8.48	8.48	8.48	8.48	9.90	9.90
8.51	8.55	8.55	8.55	8.55	8.55	8.55	8.55	8.55	8.81	8.81	8.81	8.81	10.28	10.28
11.95	11.69	11.69	11.69	11.69	11.69	11.69	11.69	11.69	12.10	12.10	12.10	12.10	14.06	14.06
11.43	11.28	11.28	11.28	11.28	11.28	11.28	11.28	11.28	11.69	11.69	11.69	11.69	13.57	13.57
7.68	7.68	7.10	7.10	7.10	7.10	7.10	7.10	7.10	7.10	7.10	7.10	7.10	7.49	7.49
54.67	59.67	59.67	59.67	59.67	59.67	59.67	59.67	59.67	61.61	61.61	61.61	61.61	71.74	71.74
45.73	39.56	39.56	39.56	39.56	39.56	39.56	39.56	39.56	41.51	41.51	41.51	41.51	47.57	47.57
20.60	20.60	20.01	20.01	20.01	20.01	20.01	20.01	20.01	20.01	20.01	20.01	20.01	21.81	21.81
16.83	16.83	16.83	16.83	16.83	16.83	16.83	16.83	16.83	16.83	16.83	16.83	16.83	18.65	18.65
19.21		19.21		19.81		21.14		22.00		22.18		22.18		

Table 1-g
TYPICAL MONTHLY FUEL BILLS FOR RESIDENTIAL CUSTOMERS
WASHINGTON, D.C.
ALL FUELS
($/Month)

	1960		1961		1962		1963		1964		1965	
	Jan.	July	Jan.	July	Jan.	July	Jan.	July	Jan.	July	Jan.	July
Non-Substitutable Base Load												
Electric	9.99	9.99	9.99	9.99	10.79	10.79	10.79	10.79	9.96	9.96	9.96	9.96
Substitutable Base Load												
Electric												
without elec. space heat.									13.89	13.89	13.89	13.89
	14.69	14.69	14.69	14.69	14.83	14.83	14.83	14.83				
with elec. space heat.									11.12	11.12	11.12	11.12
Gas	7.11	7.11	7.11	7.11	7.18	7.18	7.19	7.19	7.19	7.19	7.19	6.93
Space Heating												
Electric												
with gas sub. base load									58.91	58.91	58.91	58.91
	84.75	84.75	84.75	84.75	85.58	85.58	85.58	85.58				
with elec. sub. base load									57.67	57.67	57.67	57.67
Gas												
with elec. sub. base load	31.29	31.29	31.29	31.29	31.60	31.60	31.67	31.67	31.67	31.67	31.67	30.31
with gas sub. base load	29.88	29.88	29.88	29.88	30.17	30.17	30.17	30.17	30.17	30.24	30.24	28.88
Fuel Oil	23.58		24.71		24.85		25.41		24.65		24.58	

Note: Gas and electricity are based on prices in effect on January 1 and July 1 of the given year.

Source: American Gas Association Rate Book; FPC National Electric Rate Book.
Sources and procedures for oil estimates are described in Chapter IV.

1966		1967		1968		1969		1970		1971		1972		1973
Jan.	July	Jan.	July	Jan.	July	Jan.	July	Jan.	July	Jan.	July	Jan.	July	Jan.
9.96	9.96	9.56	9.56	9.65	9.65	9.65	9.65	9.65	9.65	9.65	9.65	9.74	9.74	12.86
13.89	13.89	13.73	13.73	13.86	13.86	13.86	13.86	13.86	13.86	13.86	13.86	14.00	14.00)	13.18
11.12	11.12	11.02	11.02	11.13	11.13	11.13	11.13	11.13	11.13	11.13	11.13	11.24	11.24)	
6.93	6.93	6.93	6.93	6.87	6.87	6.87	6.87	6.87	7.00	7.00	7.45	7.52	7.06	8.16
58.91	58.91	60.14	60.14	60.73	60.73	60.73	60.73	60.73	60.73	60.73	60.73	61.31	61.31	68.38
57.67	57.67	57.67	57.67	58.23	58.23	58.23	58.23	58.23	58.23	58.23	58.23	58.79	58.79	66.64
30.31	30.31	30.31	30.31	30.48	29.96	29.96	29.96	29.96	30.61	30.61	32.92	33.23	30.86	34.42
28.88	28.88	28.88	28.88	29.04	28.52	28.52	28.52	28.52	29.16	29.16	31.47	31.77	29.40	32.43
25.52		26.63		27.52		28.22		29.88		31.23		31.32		

Table 1-h
TYPICAL MONTHLY FUEL BILLS FOR RESIDENTIAL CUSTOMERS
MIAMI
ALL FUELS
($/Month)

	1960		1961		1962		1963		1964		1965	
	Jan.	July	Jan.	July	Jan.	July	Jan.	July	Jan.	July	Jan.	July
Non-Substitutable Base Load												
Electric	10.90	10.59	10.59	10.41	10.41	10.38	10.38	10.38	10.29	9.71	9.71	9.87
Substitutable Base Load												
Electric	14.31	13.89	13.89	13.76	13.76	13.72	13.72	13.72	13.61	13.63	13.63	13.83
Gas												
with gas space heating					)11.03	11.03	11.03	11.03	11.03	11.03	11.03	11.03
	13.53	13.53	13.53	13.53	))							
with elec. space heating					)12.95	12.95	12.95	12.95	12.95	12.95	12.95	12.95
Space Heating												
Electric												
with elec. sub. base load												
	9.55	9.28	9.28	9.19	9.19	9.16	9.16	9.16	9.08	9.10	9.10	9.24
with gas sub. base load												
Gas												
with gas sub. base load	6.74	6.74	6.74	6.74	1.76	1.76	1.76	1.76	1.76	1.76	1.76	1.76
with elec. sub. base load	8.84	8.84	8.84	8.84	8.74	8.74	8.74	8.74	8.74	8.74	8.74	8.74
Fuel Oil	--	--	--	--	--	--	--	--	--	--	--	--

Note: Gas and electricity are based on prices in effect on January 1 and July 1 of the given year.

Source: American Gas Association Rate Book; FPC National Electric Rate Book.

Gas - used year around air conditioning rates after 1961 before it became available.

1966		1967		1968		1969		1970		1971		1972		1973
Jan.	July	Jan.	July	Jan.	July	Jan.	July	Jan.	July	Jan.	July	Jan.	July	Jan.
9.87	9.92	9.91	9.91	10.00	10.09	10.09	10.09	10.09	10.06	10.06	9.89	10.14	10.51	10.85
13.84	13.90	13.89	13.89	14.03	14.15	14.15	14.14	14.14	14.10	14.10	13.87	14.23	14.75	15.21
11.03	11.03	11.03	11.36	11.47	11.40	11.38	11.49	11.49	11.48	11.48	11.93	10.91	10.91	10.91
12.95	12.95	12.95	13.34	13.47	13.40	13.35	13.50	13.50	13.48	13.48	13.94	12.62	12.62	12.62
)8.86	8.90	8.89	8.89	8.89	8.97	8.97	8.96	8.96	8.94	8.94	8.79	9.02	9.35	9.64
)														
)														
)9.24	9.29	9.28	9.28	9.28	9.36	9.36	9.36	9.36	9.33	9.62	9.17	9.41	9.76	10.06
1.76	1.76	1.76	1.81	1.81	1.71	1.70	1.78	1.78	1.77	1.77	2.05	1.76	1.76	1.76
8.74	8.74	8.74	9.00	9.00	8.96	8.94	9.02	9.02	9.01	9.01	9.28	8.76	8.76	8.76
--	--	--	--	--	--	--	--	--	--	--	--	--	--	--

Table 1-i
TYPICAL MONTHLY FUEL BILLS FOR RESIDENTIAL CUSTOMERS
NASHVILLE
ALL FUELS
($/Month)

	1960		1961		1962		1963		1964		1965	
	Jan.	July	Jan.	July	Jan.	July	Jan.	July	Jan.	July	Jan.	July
Non-Substitutable Base Load												
Electric	6.69	6.69	6.69	6.69	6.69	6.69	6.69	6.69	6.69	6.69	6.69	6.69
Substitutable Base Load												
Electric	3.30	3.30	3.30	3.30	3.30	3.30	3.30	3.30	3.30	3.30	3.30	3.30
Gas	6.32	6.72	6.72	6.72	6.72	6.72	6.72	6.32	6.32	6.32	6.32	6.32
Space Heating												
Electric												
with gas sub. base load	32.90	32.90	32.90	32.90	32.90	32.90	32.90	32.90	32.90	32.90	32.90	32.90
with elec. sub. base load	35.78	35.78	35.78	35.78	35.78	35.78	35.78	35.78	35.78	35.78	35.78	35.78
Gas												
with elec. sub. base load	23.48	25.73	25.73	25.73	25.73	25.73	25.73	23.48	23.48	23.48	23.48	23.48
with gas sub. base load	20.96	23.15	23.15	23.15	23.15	23.15	23.15	20.96	20.96	20.96	20.96	20.96
Fuel Oil	--	--	--	--	--	--	--	--	--	--	--	--

Note: Gas and electricity are based on prices in effect on January 1 and July 1 of the given year.

Source: American Gas Association Rate Book; FPC National Electric Rate Book.

1966		1967		1968		1969		1970		1971		1972		1973
Jan.	July	Jan.	July	Jan.	July	Jan.	July	Jan.	July	Jan.	July	Jan.	July	Jan.
6.69	6.69	6.69	6.69	7.00	7.00	7.00	7.20	7.20	7.41	9.20	8.99	8.99	8.99	8.99
3.30	3.30	3.30	3.30	4.25	4.25	4.25	4.64	4.64	7.24	7.95	7.95	7.95	7.95	7.95
6.32	6.32	6.32	6.32	6.32	6.32	6.32	6.32	6.32	6.32	6.32	6.89	6.89	6.89	7.39
32.90	32.90	32.90	32.90	32.97	32.97	32.97	35.31	35.31	36.72	45.15	45.15	45.15	45.15	45.15
35.78	35.78	35.78	35.78	35.99	35.99	35.99	38.33	38.33	36.41	44.78	44.78	44.78	44.78	44.78
23.48	23.48	23.48	23.48	23.48	23.48	23.48	23.48	23.48	23.48	23.48	26.52	26.52	26.52	28.98
20.96	20.96	20.96	20.96	20.96	20.96	20.96	20.96	20.96	20.96	20.96	24.00	24.00	24.00	26.41
--	--	--	--	--	--	--	--	--	--	--	--	--	--	--

Table 1-j
TYPICAL MONTHLY FUEL BILLS FOR RESIDENTIAL CUSTOMERS
CHICAGO
ALL FUELS
($/Month)

	1960		1961		1962		1963		1964		1965	
	Jan.	July	Jan.	July	Jan.	July	Jan.	July	Jan.	July	Jan.	July
Non-Substitutable Base Load												
Electric only with gas sub. base load without elec. space heat.						)11.57	11.52	11.52	11.52	11.34	11.34	11.30
	11.69	11.69	11.69	11.69	11.69	))						
with elec. space heat.						)11.30	11.24	11.24	11.24	10.55	10.55	10.51
with elec. sub. base load	8.52	8.52	8.52	8.52	8.52	7.89	7.85	7.85	7.85	7.80	7.80	7.76
Substitutable Base Load												
Electric without elec. space heating	17.06	17.06	17.06	17.06	17.06	15.55	15.64	15.64	15.64	15.50	15.50	15.42
with elec. space heating	16.01	16.01	16.01	16.01	16.01	12.05	12.13	12.13	12.13	12.33	12.33	12.33
Gas	6.03	6.02	6.02	6.02	6.11	5.93	6.01	5.86	5.92	6.06	6.05	6.04
Space Heating												
Electric	93.68	93.68	93.68	93.68	93.68	55.58	56.11	56.11	56.11	66.69	66.69	66.16
Gas with gas sub. base load	24.41	24.38	24.38	24.36	24.89	23.89	24.32	23.50	23.84	23.98	23.84	23.74
with elec. sub. base load	26.16	26.13	26.13	26.10	26.63	25.63	26.06	25.24	25.58	25.91	25.79	25.69
Fuel Oil	25.55		26.14		26.64		27.06		26.87		27.16	

Note: Gas and electricity are based on prices in effect on January 1 and July 1 of the given year.

Source: American Gas Association Rate Book; FPC National Electric Rate Book.
Sources and procedures of oil estimates are described in Chapter IV.

1966		1967		1968		1969		1970		1971		1972		1973
Jan.	July	Jan.	July	Jan.	July	Jan.	July	Jan.	July	Jan.	July	Jan.	July	Jan.
11.31	11.29	11.29	11.13	11.36	11.36	11.41	12.45	11.45	11.58	12.25	12.72	13.72	13.84	13.79
10.51	10.54	10.43	10.38	10.61	11.55	11.55	10.59	10.59	10.71	11.48	11.94	12.83	12.95	12.90
7.77	7.80	7.81	7.69	7.85	7.85	7.85	7.94	7.94	8.07	8.44	8.90	9.55	9.67	9.61
15.43	15.08	15.08	14.84	15.16	15.16	15.16	15.34	15.34	15.60	16.66	17.58	19.23	19.47	19.36
12.26	12.09	12.10	11.88	12.14	11.71	11.71	11.88	11.88	12.14	12.18	13.74	13.73	15.06	14.95
6.04	6.05	6.05	6.08	6.19	6.21	5.94	5.85	6.35	6.28	6.53	6.54	6.89	6.63	6.84
66.19	63.72	63.71	62.52	63.88	58.47	58.47	59.55	59.55	61.16	61.77	72.87	72.22	74.81	73.06
23.74	23.69	23.69	23.93	24.31	24.44	22.86	22.38	25.25	24.89	26.26	26.30	28.28	27.29	28.59
25.69	25.68	25.68	25.89	26.30	26.42	24.86	24.37	27.24	26.89	28.28	28.30	30.27	29.23	30.48
27.35		27.82		28.37		28.93		30.26		31.49		32.06		

Table 1-k
TYPICAL MONTHLY FUEL BILLS FOR RESIDENTIAL CUSTOMERS
DETROIT
ALL FUELS
($/Month)

	1960		1961		1962		1963		1964		1965	
	Jan.	July	Jan.	July	Jan.	July	Jan.	July	Jan.	July	Jan.	July
Non-Substitutable Base Load												
Electric												
with gas sub. base load												
with elec. sub. base load	11.47	11.47	11.59	11.59	11.59	11.59	11.59	11.59	11.59	11.59	11.81	11.81
Substitutable Base Load												
Electric												
without elec. space heat.											)17.42	17.42
	18.03	18.03	18.20	18.20	18.20	18.20	18.20	18.20	18.20	18.20)	)	
with elec. space heat.											)13.31	13.31
Gas												
without gas space heat.	4.91	4.91	5.35	5.35	5.35	5.35	5.24	5.24	5.24	5.24	5.24	5.24
with gas space heat.	4.66	4.66	5.08	5.08	5.08	5.08	5.02	5.02	5.02	5.02	5.01	5.01
Space Heating												
Electric	115.46	115.46	116.58	116.58	116.58	116.58	116.58	116.58	116.58	116.58	93.27	93.27
Gas												
with elec. sub. base load	22.24	22.24	23.52	23.52	23.52	23.52	23.41	23.41	23.41	23.41	23.23	23.23
with gas sub. base load	20.86	20.86	21.88	21.88	21.88	21.88	21.80	21.80	21.80	21.80	21.61	21.61
Fuel Oil	29.92		30.03		29.90		29.97		29.92		30.07	

Note: Gas and electricity are based on prices in effect on January 1 and July 1 of the given year.

Source: American Gas Association Rate Book; FPC National Electric Rate Book. Sources and procedures for oil estimates are described in Chapter IV.

1966		1967		1968		1969		1970		1971		1972		1973
Jan.	July	Jan.	July	Jan.	July	Jan.	July	Jan.	July	Jan.	July	Jan.	July	Jan.
11.82	11.82	11.82	11.82	11.82	11.82	11.82	11.82	11.82	11.82	11.82	11.94	11.94	11.94	)12.98))) 9.65
13.99	13.99	13.47	13.47	13.26	13.26	13.26	13.26	13.26	13.94	13.94	16.02	16.02	16.02)	))18.41)
13.00	13.00	13.00	13.00	12.38	12.38	12.38	12.38	12.38	12.90	12.90	14.35	14.35	14.35)	
5.24	5.24	5.24	5.24	5.24	5.24	5.24	5.24	5.24	5.24	5.24	5.61	5.61	5.61	6.16
5.01	5.01	5.01	5.00	5.00	5.00	5.00	5.00	5.00	5.00	5.00	5.00	5.00	5.40	5.96
87.44	87.44	87.44	87.44	78.70	78.70	78.70	78.70	78.70	81.61	81.61	87.44	87.44	87.44	103.02
23.23	23.23	23.23	23.12	23.12	23.12	23.12	23.12	23.12	23.12	23.12	23.60	23.60	23.60	26.37
21.61	21.61	21.61	21.48	21.49	21.49	21.49	21.49	21.49	21.49	21.49	21.59	21.59	21.59	24.21
30.57		31.67		32.33		33.35		34.74		35.84		36.32		

Table 1-1
TYPICAL MONTHLY FUEL BILLS FOR RESIDENTIAL CUSTOMERS
COLUMBUS
ALL FUELS
($/Month)

	1960		1961		1962		1963		1964		1965	
	Jan.	July	Jan.	July	Jan.	July	Jan.	July	Jan.	July	Jan.	July
Non-Substitutable Base Load												
Electric												
without elec. space heat.	11.64	11.64	11.64	11.64	11.17	11.17	11.17	11.17	10.87	10.87	10.87	10.87
with elec. space heat.	8.24	8.24	8.24	8.24	7.83	7.83	7.83	7.83	7.83	7.83	7.83	7.83
Substitutable Base Load												
Electric												
without elec. space heat.	13.39	13.39	13.39	13.39	13.39	13.39	13.39	13.39	13.39	13.39	13.39	13.39
with elec. space heat.	16.48	16.48	16.48	16.48	15.66	15.66	15.66	15.66	15.66	15.66	15.66	15.66
Gas	3.78	3.83	3.89	3.91	3.94	4.10	3.92	3.81	3.82	4.06	3.85	3.26
Space Heating												
Electric												
with gas sub. base load	89.02	89.02	89.02	89.02	83.36	83.36	83.36	83.36	83.36	83.36	83.36	83.36
with elec. sub. base load	84.90	84.90	84.90	84.90	79.24	79.24	79.24	79.24	79.24	79.24	79.24	79.24
Gas												
with elec. sub. base load	19.74	20.00	20.48	20.60	20.84	21.84	20.72	20.09	20.10	21.60	20.34	16.63
with gas sub. base load	19.00	19.25	19.75	19.88	20.12	21.12	20.00	19.38	19.38	20.88	19.62	15.92
Fuel Oil	--	--	--	--	--	--	--	--	--	--	--	--

Note: Gas and electricity are based on prices in effect on January 1 and July 1 of the given year.

Source: American Gas Association Rate Book; FPC National Electric Rate Book.

1966		1967		1968		1969		1970		1971		1972		1973
Jan.	July	Jan.	July	Jan.	July	Jan.	July	Jan.	July	Jan.	July	Jan.	July	Jan.
10.54	10.54	10.59	10.59	10.64	10.64	10.64	10.64	10.64	10.64	10.64	10.64	10.64	12.00	12.22
7.83	7.83	7.87	7.87	7.90	7.90	7.90	7.90	7.90	7.90	7.90	7.90	7.90	8.44	8.65
13.39	13.39	13.46	13.46	13.52	13.52	13.52	13.52	13.52	13.52	13.52	13.52	13.52	14.97	15.42
15.66	15.66	15.73	15.73	15.81	15.81	15.81	15.81	15.81	15.81	15.81	15.81	15.81	16.89	17.31
3.51	3.64	3.68	3.71	3.76	3.60	3.46	3.67	3.75	3.93	3.97	4.24	4.28	4.67	4.71
73.70	73.70	74.05	74.05	74.41	74.41	74.41	74.41	74.41	74.41	74.41	74.41	74.41	81.82	84.78
67.95	67.95	68.28	68.28	68.61	68.61	68.61	68.61	68.61	68.61	68.61	68.61	68.61	76.00	78.98
18.16	19.01	19.25	19.56	19.70	18.70	17.93	19.18	19.68	20.78	21.03	22.77	23.02	25.47	25.72
17.14	18.30	18.53	18.86	18.98	17.98	17.11	18.46	18.96	20.06	20.31	22.06	22.31	24.75	24.96
--	--	--	--	--	--	--	--	--	--	--	--	--	--	--

Table 1-m
TYPICAL MONTHLY FUEL BILLS FOR RESIDENTIAL CUSTOMERS
HOUSTON
ALL FUELS
($/Month)

	1960		1961		1962		1963		1964		1965	
	Jan.	July	Jan.	July	Jan.	July	Jan.	July	Jan.	July	Jan.	July
Non-Substitutable Base Load												
Electric	8.13	8.24	8.24	8.10	8.26	8.55	9.83	9.34	9.34	9.34	9.34	9.30
Substitutable Base Load												
Electric	10.80	11.00	11.00	11.14	11.36	11.52	9.49	8.51	9.34	9.34	8.51	8.43
Gas	4.41	4.41	4.41	4.41	4.50	4.50	5.12	5.12	5.12	5.12	5.12	5.12
Space Heating												
Electric												
with gas sub. base load	23.48	23.98	23.98	24.34	24.83	25.23	23.19	20.73	20.73	20.73	20.73	20.50
with elec. sub. base load	20.98	21.48	21.48	21.84	22.28	22.68	22.68	20.22	20.22	20.22	20.22	19.99
Gas												
with elec. sub. base load	8.07	8.07	8.07	8.07	8.23	8.23	9.32	9.32	9.32	9.32	9.32	9.32
with gas sub. base load	6.60	6.60	6.60	6.60	6.73	6.73	7.19	7.19	7.19	7.19	7.19	7.19
Fuel Oil	--	--	--	--	--	--	--	--	--	--	--	--

Note: Gas and electricity are based on prices in effect on January 1 and July 1 of the given year.

Source: American Gas Association Rate Book; FPC National Electric Rate Book.

1966		1967		1968		1969		1970		1971		1972		1973
Jan.	July	Jan.	July	Jan.	July	Jan.	July	Jan.	July	Jan.	July	Jan.	July	Jan.
9.30	9.33	9.33	9.33	9.42	9.42	9.56	9.56	9.56	9.56	9.64	10.37	9.33	9.33	10.58
8.43	8.49	8.49	8.49	8.57	8.57	8.69	8.69	8.69	8.69	8.80	8.80	8.82	8.82	9.20
5.12	5.12	5.12	5.12	5.17	5.17	6.12	6.12	6.12	6.12	6.15	6.15	6.62	6.62	6.62
20.50	20.68	20.68	20.68	20.88	20.88	21.18	21.18	21.18	21.18	21.44	21.44	21.49	21.49	22.44
19.99	20.17	20.17	20.17	20.36	20.36	20.66	20.66	20.66	20.66	20.92	20.92	20.97	20.97	21.92
9.32	9.32	9.32	9.32	9.41	9.41	10.89	10.89	10.89	10.89	10.94	10.94	11.95	11.95	11.95
7.19	7.19	7.19	7.19	7.26	7.26	8.11	8.11	8.11	8.11	8.15	8.15	9.11	9.11	9.11
--	--	--	--	--	--	--	--	--	--	--	--	--	--	--

Table 1-n

TYPICAL MONTHLY FUEL BILLS FOR RESIDENTIAL CUSTOMERS
NEW ORLEANS
ALL FUELS

($/Month)

	1960		1961		1962		1963		1964		1965	
	Jan.	July	Jan.	July	Jan.	July	Jan.	July	Jan.	July	Jan.	July
Non-Substitutable Base Load												
Electric												
without elec. space heating									12.32	12.32	12.26	12.26
	10.05	10.05	10.56	10.56	10.62	10.62	10.80	10.80				
with elec. space heating									8.65	8.65	8.59	6.59
Substitutable Base Load												
Electric												
without elec. space heating									12.50	12.50	12.38	12.38
	12.79	12.79	13.81	13.81	13.93	13.93	14.10	14.10				
with elec. space heating									10.77	10.77	10.65	10.65
Gas	3.12	3.22	3.22	3.26	3.44	3.44	3.44	3.44	3.44	3.44	3.44	3.44
Space Heating												
Electric	31.04	31.04	33.62	33.62	33.92	33.92	34.34	34.34	27.22	27.22	26.91	26.91
Gas												
with gas sub. base load	5.45	5.93	5.93	6.03	6.43	6.43	6.43	6.43	6.43	6.43	6.43	6.43
with elec. sub. base load	6.15	6.62	6.62	6.72	7.12	7.12	7.12	7.12	7.12	7.12	7.12	7.12
Fuel Oil	--	--	--	--	--	--	--	--	--	--	--	--

Note: Gas and electricity are based on prices in effect on January 1 and July 1 of the given year.

Source: American Gas Association Rate Book, FPC National Electric Rate Book.

1966		1967		1968		1969		1970		1971		1972		1973
Jan.	July	Jan.	July	Jan.	July	Jan.	July	Jan.	July	Jan.	July	Jan.	July	Jan.
12.16	12.16	12.14	12.14	12.49	12.49	12.49	12.49	12.61	12.53	12.53	12.53	12.63	12.76	12.76
8.49	8.49	8.47	8.47	8.71	8.71	8.71	8.71	8.80	8.69	8.69	8.69	8.82	8.95	8.95
12.18	12.18	12.14	12.14	12.49	12.49	12.49	12.49	12.61	12.40	12.40	12.40	12.66	12.91	12.91
10.44	10.44	10.40	10.40	10.71	10.71	10.71	10.71	10.81	10.60	10.60	10.60	10.85	11.11	11.11
3.44	3.44	3.44	3.44	3.54	3.54	3.54	3.54	3.57	3.57	3.57	4.10	4.10	4.39	4.39
26.40	26.40	26.23	26.23	27.01	27.01	27.01	27.01	27.26	26.79	26.79	26.79	27.43	28.07	28.07
6.43	6.43	6.43	6.43	6.62	6.62	6.62	6.62	6.68	6.68	6.68	7.89	7.89	8.54	8.55
7.12	7.12	7.12	7.12	7.33	7.33	7.33	7.33	7.40	7.40	7.40	8.61	8.61	9.26	9.28
--	--	--	--	--	--	--	--	--	--	--	--	--	--	--

Table 1-o
TYPICAL MONTHLY FUEL BILLS FOR RESIDENTIAL CUSTOMERS
TULSA
ALL FUELS
($/Month)

	1960		1961		1962		1963		1964		1965	
	Jan.	July	Jan.	July	Jan.	July	Jan.	July	Jan.	July	Jan.	July
Non-Substitutable Base Load												
Electric												
with gas sub. base load	11.67	11.67	11.72	11.74	11.74	11.84	11.84	11.84	11.81	11.81	11.93	11.93
with elec. sub. base load												
without elec. space heat.												
	8.61	8.61	8.66	8.68	8.68	8.72	8.72	8.72	8.75	8.75	8.84	8.84
with elec. space heat.												
Substitutable Base Load												
Electric												
without elec. space heating										)13.55	13.68	13.68
	8.16	8.16	8.26	8.31	8.31	8.51	8.51	8.51	8.45	)		
with elec. space heating										) 8.45	8.53	8.53
Gas	3.61	3.63	3.62	3.63	3.63	3.63	3.63	3.64	3.66	3.66	4.34	4.35
Space Heating												
Electric	41.70	41.70	42.23	42.44	42.44	43.45	43.45	43.45	43.16	43.16	43.58	43.58
Gas												
with elec. sub. base load	12.97	13.10	13.04	13.10	13.10	13.11	13.11	13.19	13.24	13.28	15.36	15.40
with gas sub. base load	11.94	12.07	12.01	12.07	12.07	12.08	12.08	12.16	12.21	12.25	13.54	13.59
Fuel Oil	--	--	--	--	--	--	--	--	--	--	--	--

Note: Gas and electricity are based on prices in effect on January 1 and July 1 of the given year.

Source: American Gas Association Rate Book; FPC National Electric Rate Book.

	1966		1967		1968		1969		1970		1971		1972		1973
	Jan.	July	Jan.	July	Jan.	July	Jan.	July	Jan.	July	Jan.	July	Jan.	July	Jan.
	11.85	11.85	11.80	11.80	11.87	11.87	11.88	11.86	11.86	11.86	11.88	11.88	12.23	12.23	11.96
	)8.76	8.76	8.71	8.71	8.52	8.52	8.53	8.77	8.77	8.77	8.79	8.79	9.11	9.11	8.84
	)														
	)														
	)8.66	8.66	8.62	8.62	8.68	8.68	8.69	8.67	8.67	8.67	8.69	8.69	9.02	9.02	8.75
	14.28	14.28	14.20	14.20	14.33	14.33	14.35	14.32	14.32	14.32	14.36	14.36	14.96	14.96	14.41
	8.35	8.35	8.28	8.28	8.40	8.40	8.42	8.39	8.39	8.39	8.44	8.44	8.98	8.98	8.43
	4.34	4.34	4.34	4.35	4.34	4.37	4.78	4.78	4.77	4.77	4.78	4.80	4.85	4.99	5.02
	42.70	42.70	42.36	42.36	42.93	42.93	43.03	42.87	42.87	42.87	43.10	43.10	45.88	45.88	42.52
	15.36	15.42	15.42	14.78	14.71	14.80	15.58	15.58	15.55	15.55	15.60	15.72	15.87	16.57	16.67
	13.54	13.61	13.61	12.50	12.43	12.52	13.77	13.77	13.74	13.74	13.79	13.91	14.04	14.74	14.84
	--	--	--	--	--	--	--	--	--	--	--	--	--	--	--

Table 1-p

TYPICAL MONTHLY FUEL BILLS FOR RESIDENTIAL CUSTOMERS
MINNEAPOLIS-ST. PAUL
ALL FUELS

($/Month)

	1960		1961		1962		1963		1964		1965	
	Jan.	July	Jan.	July	Jan.	July	Jan.	July	Jan.	July	Jan.	July
Non-Substitutable Base Load												
Electric												
with gas sub. base load	11.67	11.67	11.67	11.67	11.08	11.08	11.08	11.08	11.08	10.35	10.15	10.15
with elec. sub. base load												
without elec. space heat.	9.13	9.13	9.13	9.13)								
				)	6.91	6.91	6.91	6.91	6.91	6.70	6.60	6.60
with elec. space heat.	9.74	9.74	9.74	9.74)								
Substitutable Base Load												
Electric												
without elec. space heat.	18.27	18.27	18.27	18.27	16.80	16.80	16.80	16.80	16.80	15.94	16.24	16.24
with elec. space heat.	14.87	14.87	14.87	14.87	8.12	8.12	8.12	8.12	8.12	8.12	8.12	8.12
Gas	5.96	5.96	5.96	5.96	5.96	5.96	5.96	5.96	5.96	5.96	5.96	5.92
Space Heating												
Electric												
with gas sub. base load	100.22	100.22	100.22	100.22	88.88	88.88	88.88	88.88	88.88	77.66	77.45	77.45
with elec. sub. base load	99.56	99.56	99.56	99.56	92.88	92.88	92.88	92.88	92.88	80.70	79.59	79.59
Gas												
with elec. sub. base load	26.60	26.60	26.60	26.60	26.60	26.60	26.60	26.60	26.60	26.60	26.60	25.92
with gas sub. base load	24.48	24.48	24.48	24.48	24.48	24.48	24.48	24.48	24.48	24.48	24.48	23.72
Fuel Oil	29.32		29.01		28.68		28.68		28.85		28.72	

Note: Gas and electricity are based on prices in effect on January 1 and July 1 of the given year.

Source: American Gas Association Rate Book; FPC National Electric Rate Book.
Sources and procedures for oil estimates are described in Chapter IV.

1966		1967		1968		1969		1970		1971		1972		1973
Jan.	July	Jan.	July	Jan.	July	Jan.	July	Jan.	July	Jan.	July	Jan.	July	Jan.
10.15	10.15	10.45	10.45	10.45	10.45	1 0.45	10.45	11.03	11.03	11.03	12.16	12.49	12.54	12.64
6.60	6.60	6.80	6.80	6.80	6.80	6.80	6.80	7.16	7.16	7.16	10.39	10.70	10.75	10.86
16.24	16.24	16.73	16.73	16.73	16.73	16.73	16.73	14.84	14.84	14.84)				
										)	13.58	14.12	14.23	14.42
8.12	8.12	8.36	8.36	8.36	8.36	8.36	8.36	8.36	8.36	8.36)				
5.92	5.92	6.10	6.10	6.10	6.10	6.10	6.82	6.82	6.82	6.82	7.65	7.73	7.73	7.73
77.45	77.45	79.77	79.77	79.77	79.77	79.77	79.77	77.90	77.90	77.90	83.24	86.93	87.65	89.11
79.59	79.59	81.98	81.98	81.98	81.98	81.98	81.98	80.00	80.00	80.00	81.12	84.79	85.51	86.96
25.92	24.84	25.59	25.59	25.59	25.59	25.59	27.19	27.19	27.19	27.19	29.57	29.86	29.86	29.86
23.72	22.44	23.11	23.11	23.11	23.11	23.11	24.16	24.16	24.16	24.16	26.01	26.26	26.26	26.26
28.68		30.20		31.29		32.21		33.66		34.66		34.99		

Table 1-q

TYPICAL MONTHLY FUEL BILLS FOR RESIDENTIAL CUSTOMERS
OMAHA
ALL FUELS
($/Month)

	1960		1961		1962		1963		1964		1965	
	Jan.	July	Jan.	July	Jan.	July	Jan.	July	Jan.	July	Jan.	July
Non-Substitutable Base Load												
Electric												
publicly owned												
with gas sub. base load											)8.05	8.05
with elec. sub. base load without elec. space heat.	8.72	8.72	8.72	8.72	8.72	8.72	8.72	8.72	8.72	8.72	)7.30	7.30
with elec. space heat.											)7.00	7.00
privately owned												
with gas sub. base load	10.77	10.83	10.83	10.78	10.78	10.83	10.78)					
with elec. sub.								)11.49	11.42	11.42	11.42	11.42
base load	9.70	9.77	9.77	9.71	9.71	9.76	9.71)					
Substitutable Base Load												
Electric												
publicly owned												
without elec. space heat.			)12.00	12.00	12.00	12.00	12.00	12.00	12.00	12.00	8.75	8.75
with elec.	10.00	10.00)										
space heat.			)10.00	10.00	10.00	10.00	10.00	10.00	10.00	10.00	7.60	7.60
privately owned												
without elec. space heat.												
with elec. space heat.	13.55	13.93	13.93	13.81	13.81	13.93	13.81	12.36	12.24	12.24	12.24	12.24
Gas	4.60	4.60	4.60	4.60	4.60	4.60	4.60	4.38	4.38	4.38	3.90	3.90
Space Heating												
Electric												
publicly owned												
with gas sub. base load			)56.29	56.29	56.29	56.29	56.29	56.29	56.29	56.29	56.29	56.29
with elec. sub.	81.43	81.43)										
base load			)54.29	54.29	54.29	54.29	54.29	54.29	54.29	54.29	48.86	48.86
privately owned												
with gas sub. base load								)74.45	73.62	73.62	73.62	73.62
with elec. sub.	94.96	95.80	95.80	94.97	94.79	95.80	94.97)					
base load								)72.82	71.99	71.99	71.99	71.99
Gas												
with elec. sub. base load	18.51	18.51	18.51	18.51	18.51	18.51	18.51	17.25	17.25	17.25	15.35	15.35
with gas sub. base load	16.47	16.47	16.47	16.47	16.47	16.47	16.47	15.23	15.23	15.23	13.55	13.55
Fuel Oil	--	--	--	--	--	--	--	--	--	--	--	--

Note: Gas and electricity are based on prices in effect on January 1 and July 1 of the given year.

Source: American Gas Association Rate Book; FPC National Electric Rate Book.

	1966		1967		1968		1969		1970		1971		1972		1973
	Jan.	July	Jan.	July	Jan.	July	Jan.	July	Jan.	July	Jan.	July	Jan.	July	Jan.
	8.05	8.05	8.05	8.05	8.25	8.25	8.25	8.25	8.25	8.25	9.42	9.42	9.42	9.42	9.42
	7.30	7.30	7.30	7.30	7.48	7.48	7.48	7.48	7.48	7.48	8.49	8.49	8.49	8.49	8.49
	7.00	7.00	7.00	7.00	7.17	7.17	7.17	7.17	7.17	7.17	7.76	7.76	7.76	7.76	7.76
	11.42	11.49	11.49	11.42	11.54	11.54	11.54	11.54	11.33	11.33	11.33	12.69	12.63	12.79	12.84
	8.75	8.75	8.75	8.75	8.97	8.97	8.97	8.97	8.97	8.97	11.70	11.70	11.70	11.70	11.70
	7.60	7.60	7.60	7.60	7.79	7.79	7.79	7.79	7.79	7.79	7.87	7.87	7.87	7.87	7.87
	12.24	12.36	12.36	12.24	12.36	12.36	12.36	12.36	12.36	12.36	12.36)	)16.10	16.10	16.92	17.03
												)15.69	15.57	15.89	16.00
	4.38	4.38	4.06	4.06	4.16	4.16	4.21	4.62	4.62	4.67	4.67	5.06	5.06	5.06	5.06
	56.29	56.29	56.29	56.29	57.70	57.70	57.70	57.70	57.70	57.70	58.67	58.67	58.67	58.67	58.67
	48.86	48.86	48.86	48.86	50.08	50.08	50.08	50.08	50.08	50.08	50.57	50.57	50.57	50.57	50.57
	73.62	74.45	74.45	73.62	74.35	74.35	74.35	74.35	74.35	74.35	74.35	81.38	80.66	82.83	83.56
	71.99	72.82	72.82	71.99	72.70	72.70	72.70	72.70	72.70	72.70	72.70	77.05	76.32	78.51	79.24
	17.25	17.25	15.67	15.67	16.06	16.06	16.24	17.27	17.27	17.64	17.64	19.52	19.52	19.52	19.52
	15.23	15.23	13.81	13.81	14.15	14.15	14.31	15.01	15.01	15.39	15.39	17.18	17.18	17.18	17.18
	--	--	--	--	--	--	--	--	--	--	--	--	--	--	--

Table 1-r
TYPICAL MONTHLY FUEL BILLS FOR RESIDENTIAL CUSTOMERS
ST. LOUIS
ALL FUELS
($/Month)

	1960		1961		1962		1963		1964		1965	
	Jan.	July	Jan.	July	Jan.	July	Jan.	July	Jan.	July	Jan.	July
Non-Substitutable Base Load												
Electric												
with gas sub. base load	10.46	10.46	10.46	10.46	10.56	10.56	10.56	10.56	10.56	10.66	10.66	10.66
with elec. sub. base load	8.77	8.77	8.77	8.77	8.86	8.86	8.86	8.86	8.86	9.48	9.48	9.48
Substitutable Base Load												
Electric	11.73	11.73	11.73	11.73	11.85	11.85	11.85	11.85	11.85	12.69	12.69	12.69
Gas	6.13	6.30	6.30	6.30	6.37	6.37	6.32	6.37	6.37	6.45	6.33	6.33
Space Heating												
Electric												
with gas sub. base load	55.02	55.02	55.02	55.02	55.56	55.56	55.56	55.56	55.56	51.21	51.21	51.21
with elec. sub. base load	52.06	52.06	52.06	52.06	52.57	52.57	52.57	52.57	52.57	46.40	46.40	46.40
Gas												
with elec. sub. base load	16.49	17.77	17.77	17.77	17.94	17.94	17.77	17.95	17.95	18.13	17.72	17.72
with gas sub. base load	13.50	14.61	14.61	14.61	14.75	14.75	14.60	14.78	14.78	14.92	14.52	14.52
Fuel Oil	19.94		19.99		20.22		20.36		20.53		20.59	

Note: Gas and electricity are based on prices in effect on January 1 and July 1 of the given year.

Source: American Gas Association Rate Book; FPC National Electric Rate Book. Sources and procedures for oil estimates are described in Chapter IV.

	1966		1967		1968		1969		1970		1971		1972		1973
	Jan.	July	Jan.	July	Jan.	July	Jan.	July	Jan.	July	Jan.	July	Jan.	July	Jan.
	10.66	10.66	10.66	10.66	10.25	10.36	10.36	10.36	11.01	11.01	11.01	11.01	11.50	11.50	12.36
	9.48	9.48	9.48	9.48	9.12	9.22	9.22	9.22	9.80	9.80	9.80	9.80	10.23	10.23	10.89
	12.69	12.69	12.69	12.69	12.20	12.11	12.11	12.11	13.08	13.08	13.08	13.08	13.67	13.67	16.22
	6.33	6.33	6.33	6.33	6.33	6.33	6.39	6.39	6.57	6.85	6.85	7.17	7.28	7.28	7.57
	51.21	51.21	51.21	51.21	49.24	36.29	36.29	36.29	37.19	37.19	37.19	37.19	38.84	38.84	41.13
	46.40	46.40	46.40	46.40	44.62	30.17	30.17	30.17	30.17	30.17	30.17	30.17	31.51	31.51	31.51
	17.72	17.69	17.69	17.69	17.69	17.69	17.86	17.86	18.20	19.35	19.35	20.75	21.08	21.08	22.42
	14.52	14.49	14.49	14.49	14.49	14.49	14.62	14.62	14.81	15.95	15.95	17.35	17.66	17.66	18.99
	20.77		21.33		21.98		22.56		23.47		24.83		25.29		

Table 1-s

TYPICAL MONTHLY FUEL BILLS FOR RESIDENTIAL CUSTOMERS
DENVER
ALL FUELS
($/Month)

	1960		1961		1962		1963		1964		1965	
	Jan.	July	Jan.	July	Jan.	July	Jan.	July	Jan.	July	Jan.	July
Non-Substitutable Base Load												
Electric												
without elec. sub. base load and elec. space heat.								10.61	10.61	10.48	10.58	10.58
with elec. sub. base load and elec. space heat.	9.73	9.73	10.61	10.61	10.61	10.61	10.61	10.61	10.61	10.61	10.71	10.71
Substitutable Base Load												
Electric												
without elec. space heat.								17.14	17.14	17.14	17.30	17.30
with elec. space heat.	16.32	16.32	17.14	17.14	17.14	17.14	17.14	12.24	12.24	12.24	12.36	12.36
Gas	2.85	3.00	3.48	3.48	3.48	3.48	3.48	3.48	3.48	3.48	3.51	3.51
Space Heating												
Electric												
with elec. sub. base load								75.67	75.67	75.67	76.42	76.42
with gas sub. base load	100.90	100.90	105.95	105.95	105.95	105.95	105.95	108.25	108.25	105.95	106.99	106.99
Gas												
with gas sub. base load	6.76	7.60	10.23	10.23	10.23	10.23	10.23	10.23	10.23	10.23	10.33	10.33
with elec. sub. base load	8.42	9.25	11.91	11.91	11.91	11.91	11.91	11.91	11.91	11.91	12.03	12.03
Fuel Oil	--	--	--	--	--	--	--	--	--	--	--	--

Note: Gas and electricity are based on prices in effect on January 1 and July 1 of the given year.

Source: American Gas Association Rate Book; FPC National Electric Rate Book.

1966		1967		1968		1969		1970		1971		1972		1973
Jan.	July	Jan.	July	Jan.	July	Jan.	July	Jan.	July	Jan.	July	Jan.	July	Jan.
10.58	10.58	10.89	10.89	10.89	10.89	10.89	10.89	10.89	10.89	10.89	10.89	10.89	11.37	11.26
10.71	10.71	11.02	11.02	11.02	11.02	11.02	11.02	11.02	11.02	11.02	11.02	11.02	11.17	11.13
17.30	17.30	17.81	17.81	15.90	15.90	15.90	15.90	15.90	15.90	15.90	15.90	15.90	18.13	17.94
12.36	12.36	12.72	12.72	12.08	12.08	12.08	12.08	12.08	12.08	12.08	12.08	12.08	13.55	13.36
3.51	3.51	3.61	3.61	3.61	3.61	3.61	3.57	3.59	3.59	3.99	3.99	4.00	4.00	4.00
76.42	76.42	78.64	78.64	62.91	62.91	62.91	62.91	62.91	62.91	62.91	62.91	62.91	79.82	79.28
106.99	106.99	110.10	68.93	68.93	68.93	68.93	68.93	68.93	68.93	68.93	68.93	68.93	80.46	78.64
10.33	10.33	10.63	10.63	10.63	10.63	10.63	10.63	10.71	10.71	11.30	11.30	11.42	11.42	11.42
12.03	12.03	12.38	12.38	12.38	12.38	12.38	12.34	12.42	13.42	13.42	13.42	13.54	13.54	13.54
--	--	--	--	--	--	--	--	--	--	--	--	--	--	--

Table 1-t

TYPICAL MONTHLY FUEL BILLS FOR RESIDENTIAL CUSTOMERS
PHOENIX
ALL FUELS
($/Month)

	1960		1961		1962		1963		1964		1965	
	Jan.	July	Jan.	July	Jan.	July	Jan.	July	Jan.	July	Jan.	July
Non-Substitutable Base Load												
Electric	10.29	10.29	10.29	10.29	10.29	10.29	10.29	10.29	10.57	10.57	10.57	10.57
Substitutable Base Load												
Electric	14.86	14.86	14.97	14.97	14.97	14.97	14.97	14.97	15.07	15.07	15.07	15.07
Gas	5.06	5.06	6.04	6.04	6.02	6.02	5.96	5.96	5.88	6.01	5.93	5.93
Space Heating												
Electric												
with gas sub. base load												
with elec. sub. base load	29.80	29.80	30.23	30.23	30.23	20.23	30.23	30.23	30.23	30.23	30.23	30.23
Gas												
with elec. sub. base load	7.50	7.50	8.99	8.99	8.95	8.95	8.84	8.84	8.71	8.93	8.77	8.77
with gas sub. base load	5.23	5.23	6.33	6.33	6.29	6.29	6.18	6.18	6.05	6.26	6.10	6.10
Fuel Oil	--	--	--	--	--	--	--	--	--	--	--	--

1/ Winter rates become available.

Note: Gas and electricity are based on prices in effect on January 1 and July 1 of the given year.

Source: American Gas Association Rate Book; FPC National Electric Rate Book.

1966		1967		1968		1969		1970		1971		1972		1973
Jan.	July	Jan.	July	Jan.	July	Jan.	July	Jan.	July	Jan.	July	Jan.	July	Jan.
10.58	10.58	10.61	10.61	10.85	10.85	10.98[1/]	11.00	11.00	11.00	11.04	11.04	11.12	12.41	12.41
15.10	15.10	15.16	15.16	15.49	15.49	9.77[1/]	9.82	9.82	9.82	9.91	9.91	10.05	11.82	11.82
6.04	6.04	6.04	6.03	6.02	6.05	6.14	6.13	6.18	6.18	6.18	6.48	6.48	6.85	6.85
30.28	30.28	30.41	30.41	31.07	31.07	17.35[1/]	17.41	17.41	17.41	17.58	17.58	17.88	)20.36)18.66	20.36 18.66
8.94	8.94	8.94	8.93	8.88	8.94	9.07	9.03	9.18	9.18	9.18	9.71	9.71	10.27	10.27
6.23	6.23	6.23	6.22	6.12	6.19	6.28	6.25	6.43	6.43	6.43	6.96	6.96	7.43	7.43
--	--	--	--	--	--	--	--	--	--	--	--	--	--	--

Table 1-u
TYPICAL MONTHLY FUEL BILLS FOR RESIDENTIAL CUSTOMERS
SALT LAKE CITY
ALL FUELS
($/Month)

	1960		1961		1962		1963		1964		1965	
	Jan.	July	Jan.	July	Jan.	July	Jan.	July	Jan.	July	Jan.	July
Non-Substitutable Base Load												
Electric												
with gas sub. base load	9.23	9.23	9.27	9.27	9.27	9.27	10.00	10.00	10.00	10.00	10.00	10.00
with elec. sub. base load												
without elec. space heating										)8.55	8.55	8.55
with elec.	7.94	7.94	7.98	7.98	7.98	7.98	8.55	8.55	8.55	)		
space heating										)8.55[1/]	8.55	8.55
Substitutable Base Load												
Electric												
without elec. space heating										)11.64	11.64	11.64
with elec.	12.71	12.71	12.77	12.77	12.77	12.77	11.64	11.64	11.64	)		
space heating										) 9.52	9.52	9.52
Gas	3.83	3.92	3.94	3.94	3.90	3.90	3.92	3.92	3.92	3.87	3.87	3.87
Space Heating												
Electric	91.49	91.49	91.94	91.94	91.94	91.94	98.35	98.35	98.35	83.45[1/]	83.45	83.45
Gas												
with elec. sub. base load	13.79	14.42	14.49	14.49	14.35	14.35	14.42	14.42	14.42	14.28	14.28	14.28
with gas sub. base load	11.74	12.39	12.45	12.45	12.33	12.33	12.39	12.39	12.39	12.27	12.27	12.27
Fuel Oil	--	--	--	--	--	--	--	--	--	--	--	--

1/ All electric home rate available to present.

Note: Gas and electricity are based on prices in effect on January 1 and July 1 of the given year.

Source: American Gas Association Rate Book; FPC National Electric Rate Book.

1966		1967		1968		1969		1970		1971		1972		1973
Jan.	July	Jan.	July	Jan.	July	Jan.	July	Jan.	July	Jan.	July	Jan.	July	Jan.
10.00	10.00	10.00	10.00	10.00	10.00	10.09	10.09	10.09	10.09	10.09	10.09	10.09	9.71	10.94
8.55	8.55	8.55	8.55	8.55	8.55	8.63	8.63	8.63	8.63	8.63	8.63	8.63	8.30	9.26
8.55	8.55	8.55	8.55	8.55	8.55	8.31	8.31	8.31	8.31	8.31	8.31	8.31	7.99	8.97
11.64	11.64	11.64	11.64	11.64	11.64	11.76	11.76	11.76	11.76	11.76	11.76	11.76	11.31	13.44
9.52	9.52	9.52	9.52	9.52	9.52	9.20	9.20	9.20	9.20	9.20	9.20	9.20	8.84	10.03
3.87	3.87	3.87	3.87	3.87	3.87	7.50	7.50	7.36	7.51	7.51	7.51	7.51	7.51	7.51
83.45	83.45	83.45	83.45	83.45	83.45	66.20	66.20	66.20	66.20	66.20	66.20	66.20	63.67	72.23
14.28	14.28	14.28	14.28	14.28	14.28	19.04	19.04	18.67	19.04	19.04	19.04	19.04	19.90	19.90
12.27	12.27	12.27	12.27	12.27	12.27	13.56	13.56	13.30	13.56	13.56	13.56	13.56	14.58	14.58
--	--	--	--	--	--	--	--	--	--	--	--	--	--	--

Table 1-v
TYPICAL MONTHLY FUEL BILLS FOR RESIDENTIAL CUSTOMERS
LOS ANGELES
ALL FUELS
($/Month)

	1960		1961		1962		1963		1964		1965	
	Jan.	July	Jan.	July	Jan.	July	Jan.	July	Jan.	July	Jan.	July
Non-Substitutable Base Load												
Electric												
Privately-Owned Utility												
with gas sub. base load	8.97	8.97	8.97	8.97	8.97	8.97	8.97	8.87	8.87	8.77	8.55	8.55
with elec. sub. base load	8.38	8.38	8.38	8.38	8.38	8.38	8.38	8.28	8.28	8.18	7.96	7.96
Publicly-Owned Utility												
with gas sub. base load	8.03	8.03	8.03	8.03	8.20	8.20	8.20	8.16	8.16	8.08	8.08	7.71
with elec. sub. base load	7.42	7.42	7.42	7.42	7.64	7.64	7.64	7.54	7.54	7.46	7.46	7.12
Substitutable Base Load												
Electric												
Privately-Owned Utility	10.15	10.15	10.15	10.15	10.15	10.15	10.15	10.15	10.15	10.73	10.73	10.73
Publicly-Owned Utility	10.18	10.18	10.18	10.18	10.51	10.51	10.51	10.42	10.42	10.26	10.26	9.79
Gas	5.87	5.87	5.87	5.87	5.87	5.87	5.87	6.03	6.03	5.93	5.93	5.93
Space Heating												
Electric												
Privately-Owned Utility												
with gas sub. base load												
with elec. sub. base load	31.06	31.06	31.06	31.06	31.06	31.06	31.06	31.06	31.06	31.06	31.06	31.06
Publicly-Owned Utility												
with gas sub. base load												
with elec. sub. base load	29.07	29.07	29.07	29.07	29.95	29.95	29.95	29.73	29.73	29.28	29.28	27.75
Gas												
with gas sub. base load	8.63	8.63	8.63	8.63	8.63	8.63	8.63	8.93	8.93	8.71	8.71	8.71
with elec. sub. base load	11.11	11.11	11.11	11.11	11.11	11.11	11.11	11.41	11.41	11.19	11.19	11.19
Fuel Oil	--	--	--	--	--	--	--	--	--	--	--	--

Note: Gas and electricity are based on prices in effect on January 1 and July 1 of the given year.

Source: American Gas Association Rate Book; FPC National Electric Rate Book.

1966		1967		1968		1969		1970		1971		1972		1973
Jan.	July	Jan.	July	Jan.	July	Jan.	July	Jan.	July	Jan.	July	Jan.	July	Jan.
8.55	8.55	8.55	8.55	8.72	8.72	8.72	8.72	10.14	10.14	10.14	10.14	11.91	12.04)	)12.35
7.96	7.96	7.96	7.96	8.12	8.12	8.12	8.12	10.04	10.04	10.04	10.04	11.59 (handwritten: 9)	11.72)	
7.99	7.99	8.16	8.16	8.16	8.16	8.16	8.16	8.16	8.49)	) 8.97	8.97	9.30	11.38	11.43
7.37	7.37	7.53	7.53	7.53	7.53	7.53	7.53	7.53	7.84)					
10.73	10.73	10.73	10.73	10.94	10.94	10.94	10.94	11.78	11.78	11.78	11.78	12.87	13.76	13.44
10.09	10.09	10.36	10.36	10.36	10.36	10.36	10.36	10.36	10.75	11.07	11.07	11.75	14.58	14.66
5.33	5.33	5.64	5.64	5.64	5.64	5.64	5.77	5.77	5.77	6.53	6.53	6.51	6.51	6.96
31.06	31.06	31.06	31.06	31.67	31.67	31.67	31.67)	) 27.69	27.69	27.69	27.69	31.12	31.83	33.51
								) 28.76	28.76	28.76	28.76	29.54	30.25	31.92
28.84	28.84	29.36	29.36	29.36	29.36	29.36	29.36	29.36	30.68)	)28.14	28.14	29.98	37.27	37.49
										)27.16	27.16	29.00	36.09	36.31
7.96	7.96	8.46	8.46	8.46	8.46	8.46	8.64	8.64	8.64	8.34	8.34	8.31	8.31	8.91
10.54	10.54	11.15	11.15	11.15	11.15	11.15	11.33	11.33	11.33	11.47	11.47	11.44	11.44	12.24
--	--	--	--	--	--	--	--	--	--	--	--	--	--	--

Table 1-w

TYPICAL MONTHLY FUEL BILLS FOR RESIDENTIAL CUSTOMERS
SAN FRANCISCO
ALL FUELS

($/Month)

	1960		1961		1962		1963		1964		1965	
	Jan.	July	Jan.	July	Jan.	July	Jan.	July	Jan.	July	Jan.	July
Non-Substitutable Base Load												
Electric	8.96	8.96	8.96	8.96	8.96	8.96	8.96	8.88	8.88	8.36	8.36	8.36
Substitutable Base Load												
Electric	10.55	10.55	10.55	10.55	10.55	10.55	10.55	10.55	10.55	10.55	10.55	10.55
Gas	3.48	3.48	3.48	3.48	3.48	3.48	3.48	3.78	3.76	3.76	3.76	3.76
Space Heating												
Electric												
with elec. sub. base load with gas sub. base load	50.72	50.72	50.72	50.72	50.72	50.72	50.72	50.72	50.72	50.72	50.72	50.72
Gas												
with gas sub. base load	11.20	11.20	11.20	11.20	11.20	11.20	11.20	11.57	11.45	11.45	11.45	11.45
with elec. sub. base load	12.13	12.13	12.13	12.13	12.13	12.13	12.13	12.71	12.61	12.61	12.61	12.61
Fuel Oil	--	--	--	--	--	--	--	--	--	--	--	--

Note: Gas and electricity are based on prices in effect on January 1 and July 1 of the given year.

Source: American Gas Association _Rate Book_; FPC _National Electric Rate Book_.

1966		1967		1968		1969		1970		1971		1972		1973
Jan.	July	Jan.	July	Jan.	July	Jan.	July	Jan.	July	Jan.	July	Jan.	July	Jan.
8.36	8.36	8.53	8.53	8.53	8.53	8.53	8.53	8.85	8.85	8.85	8.85	9.80	9.80	9.80
10.55	10.55	10.75	10.75	10.75	10.75	10.75	10.75	11.16	11.16	11.16	11.16	12.87	12.87	12.87
3.76	3.76	3.77	3.77	3.77	3.77	3.77	3.83	3.98	4.40	4.40	4.40	4.39	4.39	4.39
												)55.60	55.60	55.60
50.72	50.72	51.70	51.70	51.70	51.70	51.70	51.70	53.67	53.67	53.67	53.67)	)		
												)56.91	56.91	56.91
11.45	11.45	11.38	11.38	11.38	11.38	11.38	11.64	12.09	13.49	13.49	13.49	13.46	13.46	13.46
12.61	12.61	12.57	12.57	12.57	12.57	12.57	12.84	13.33	14.88	14.88	14.88	14.84	14.84	14.84
--	--	--	--	--	--	--	--	--	--	--	--	--	--	--

Table 1-x
TYPICAL MONTHLY FUEL BILLS FOR RESIDENTIAL CUSTOMERS
SEATTLE
ALL FUELS
($/Month)

	1960		1961		1962		1963		1964		1965	
	Jan.	July	Jan.	July	Jan.	July	Jan.	July	Jan.	July	Jan.	July
Non-Substitutable Base Load												
Electric												
publicly owned without elec. space heat.	5.74	5.74	5.74	5.74	5.74	5.74	5.74	5.74	5.74	5.74	5.75	5.75
with elec. space heat.	4.16	4.16	4.16	4.16	4.16	4.16	4.16	4.16	4.16	4.16	3.96	3.96
privately owned with gas sub. base load	8.03	8.46	8.46	8.46	8.46	8.46	8.46	8.46	8.78	8.78	8.79	8.79
with elec. sub. base load	5.63	5.63	5.63	5.63	5.63	6.08	6.08	6.08	6.33	6.33	6.35	6.35
Substitutable Base Load												
Electric												
publicly owned	6.22	6.22	6.22	6.22	6.22	6.22	6.22	6.22	6.22	6.22	6.17	6.17
privately owned	6.45	6.45	6.45	6.45	6.45	7.80	7.80	7.80	7.76	7.76	7.77	7.77
Gas												
without gas space heat.	12.95	12.95	12.95	12.95	12.95	12.48	12.48	12.48	12.48	12.48	12.50	12.50
with gas space heat.	8.04	8.04	8.04	8.04	8.04	8.04	8.04	7.57	7.57	7.57	7.00	7.00
Space Heating												
Electric												
publicly owned	38.85	38.05	38.05	38.05	38.05	38.05	38.05	38.05	38.05	38.85	38.93	38.93
privately owned	46.18	46.18	46.18	46.18	46.18	55.41	55.41	55.41	52.31	52.31	52.41	52.41
Gas												
with elec. sub. base load	31.35	31.35	31.35	31.35	31.35	31.35	31.35	28.92	28.92	28.92	28.78	28.78
with gas sub. base load	30.30	30.30	30.30	30.30	30.30	30.30	30.30	26.48	26.48	26.48	26.35	26.35
Fuel Oil	26.62		26.25		27.19		27.85		27.82		27.74	

1/ New rate available, using same rate as before would have risen to $12.33.

Note: Gas and electricity are based on prices in effect on January 1 and July 1 of the given year.

Source: American Gas Association Rate Book; FPC National Electric Rate Book.
Sources and procedures for oil estimates are described in Chapter IV.

1966		1967		1968		1969		1970		1971		1972		1973
n.	July	Jan.	July	Jan.	July	Jan.	July	Jan.	July	Jan.	July	Jan.	July	Jan.
.75	5.75	5.75	5.75	5.77	5.77	5.77	5.77	5.80	5.80	5.80	5.80	6.49	6.49	6.49
.96	3.96	3.79	3.79	3.80	3.80	3.80	3.80	3.82	3.82	3.82	3.82	4.20	4.20	4.20
.79	8.79	8.80	8.80	8.83	8.83	8.83	8.83	8.83	8.83	8.83	8.83	8.83	9.29	9.29
.35	6.35	6.35	6.35	6.36	6.36	6.36	6.36	6.36	6.36	6.36	6.36	6.36	6.99	6.99
.17	6.17	6.17	6.17	6.19	6.19	6.19	6.19	6.22	6.22	6.22	6.22	6.56	6.56	6.56
.77	7.77	7.79	7.79	7.82	7.82	7.82	7.82	7.82	7.82	7.82	7.82	7.82	8.40	8.40
.63	12.63	12.63	12.63	12.67	12.67	12.78	12.78	12.78	11.80	11.80	11.80	11.80	11.80	9.57 1/
.06	6.84	6.84	6.84	6.85	6.85	6.92	6.92	6.92	7.22	7.22	7.22	7.22	7.22	8.42
.93	38.93	38.93	38.93	39.04	39.04	39.04	39.04	39.23	39.23	39.23	39.23	40.40	40.40	40.40
.41	52.41	52.42	52.42	52.57	52.57	52.57	52.57	52.57	52.57	52.57	52.57	52.57	55.68	55.68
.05	29.05	29.05	29.05	29.13	29.13	29.41	29.41	29.41	29.85	29.85	29.85	29.85	29.85)	) 27.41
.60	25.02	25.02	24.62	24.69	24.69	24.92	24.92	24.92	24.92	24.92	24.92	24.92	24.92)	)
28.00		28.78		29.59		30.54		31.87		33.74		33.92		

Table 2
Sheet 1 of 4
TYPICAL MONTHLY BILLS PAID BY RESIDENTIAL CUSTOMERS FOR CENTRAL AIR CONDITIONING
GAS AND ELECTRICITY
($/Month)

Census Region SMSA	1960 Jan.	1960 July	1961 Jan.	1961 July	1962 Jan.	1962 July	1963 Jan.	1963 July	1964 Jan.	1964 July
New England										
Boston, Mass.										
Gas										
with gas sub. base load	16.19	16.19	15.55	15.55	14.64	14.64	14.64	14.64	15.16	15.09
with elec. sub. base load	20.80	20.80	21.11	21.11	20.20	20.20	20.20	20.20	20.72	20.65
Electric										
with gas sub. base load										
without elec. space heating								(17.61	17.57	17.57
with elec. space heating								(		
with elec. sub. base load	18.65	17.57	17.80	17.80	17.72	17.72	17.61	(		
without elec. space heating								(13.00	12.96	12.96
with elec. space heating								(10.69	10.65	10.65
Middle Atlantic										
Albany, N.Y.										
Gas										
with gas sub. base load	11.76	11.76	10.93	10.98	10.98	10.98	10.76	10.64	10.80	10.72
with elec. sub. base load	13.88	13.88	14.68	14.68	14.68	14.68	14.50	14.37	14.53	14.46
Electric	8.49	8.49	8.49	8.49	8.79	8.79	8.87	8.87	8.87	9.22
New York, N.Y.										
Gas										
with gas sub. base load	18.32	18.32	16.53	16.75	16.94	16.76	16.96	16.32	16.34	16.82
with elec sub. base load	22.23	20.58	20.47	20.68	20.88	20.69	20.94	20.30	20.32	20.77
Electric										
without elec. space heating										
with gas sub. base load	17.00	16.77	16.77	16.77	17.01	17.01	(17.18	16.84	16.84	16.84
with elec. sub. base load							(17.05	16.72	16.72	16.72
with elec. space heating	--	--	--	--	--	--	11.97 1/	11.87	11.87	12.12
Pittsburgh, Pa.										
Gas										
with gas sub. base load	9.58	9.58	9.58	9.47	9.47	9.96	9.91	9.88	9.87	10.10
with elec. sub. base load	11.60	11.60	11.60	10.48	10.48	11.36	11.31	11.28	11.27	11.49
Electric										
without elec. space heating	17.99	17.99	17.99	17.99	17.99	17.99	17.99	17.99	17.99	17.99
with elec. space heating										
with gas sub. base load	14.41	14.41	14.41	14.41	14.41	14.41	14.41	14.41	14.41	11.60
with elec. sub. base load	13.50	13.50	13.50	13.50	13.50	13.50	13.50	13.50	13.50	10.79
South Atlantic										
Atlanta, Ga.										
Gas										
with gas sub. base load	24.15	24.15	25.74	25.74	25.74	25.74	25.74	18.93	18.93	18.93
with elec. sub. base load	25.73	25.73	26.58	26.58	26.58	26.58	27.38	21.31	21.31	22.06
Electric										
without elec. space heating										
with gas sub. base load	21.94	21.94	18.28	18.28	18.28	18.28	18.28	18.28	18.28	18.28
with elec. sub. base load										
with elec. space heating	32.91	32.91	29.25	29.25	29.25	29.25	29.25	29.25	29.25	29.25
Charlotte										
Gas										
with gas sub. base load	21.85	21.85	21.85	21.85	21.85	21.85	21.85	21.85	21.85	21.85
with elec. sub. base load	26.56	26.56	26.56	26.68	26.68	26.68	26.48	26.48	26.48	26.48
Electric										
without elec. space heating										
with gas sub. base load	36.57	36.57	36.57	36.57	36.57	36.57	36.57	(31.81	31.81	30.75
with elec. sub. base load								(27.43	27.43	25.75
with elec. space heating	27.43	27.43	27.43	27.43	27.43	27.43	27.43	22.86	22.86	21.94
Washington, D.C.										
Gas										
with gas sub. base load	25.28	25.28	25.63	25.63	25.63	25.63	25.63	25.63	25.63	25.63
with elec. sub. base load	29.03	29.03	29.31	29.31	29.31	29.31	29.31	29.31	29.31	29.31
Electric										
with gas sub. base load	32.59	32.59	32.59	32.59	32.91	32.91	32.91	32.91	32.91	32.91
with elec. sub. base load										
Miami, Fla.										
Gas										
without gas space heating										
with gas sub. base load	68.24	68.24	68.24	68.24	67.51	64.90	64.90	64.90	64.90	64.90
with elec. sub. base load	74.68	74.68	74.68	74.68	73.94	71.28	71.28	71.28	71.28	71.28
with gas space heating										
with gas sub. base load						4/	24.49	24.49	24.49	24.49
with elec. sub. base load						4/	32.63	32.63	32.63	32.63
Electric										
with gas sub. base load	35.47	34.45	34.45	35.16	35.16	34.04	34.04	34.04	33.74	(33.65
with elec. sub. base load										(33.36

1965		1966		1967		1968		1969		1970		1971		1972		1973
Jan.	July	Jan.	July	Jan.	July	Jan.	July	Jan.	July	Jan.	July	Jan.	July	Jan.	July	Jan.
15.06	15.06	15.06	15.06	14.94	15.01	15.01	15.01	15.01	15.76	15.76	15.74	15.74	14.43 1/	14.43	14.43	16.32
20.62	20.62	20.62	20.62	20.50	24.57	24.57	24.57	24.57	25.80	25.80	22.25	22.25	21.89	21.89	21.89	23.98
15.84	15.84	15.80	15.80	(14.62	14.62	14.62	14.62	14.62	14.62	14.62	14.62	14.62	16.93	17.31	17.07	16.55
				(9.06	9.06	9.06	9.06	9.06	9.06	9.06	9.06	9.06	11.34	11.72	11.48	10.96
12.09	12.09	12.05	12.05	10.90	10.90	10.90	10.90	10.90	10.90	10.90	10.90	10.90	13.18	13.59	13.32	12.80
10.37	10.37	10.33	10.33	8.89	8.89	8.89	8.89	8.89	8.89	8.89	8.89	8.89	11.17	11.55	11.31	10.79
10.85	10.85	10.85	10.85	10.94	10.94	10.96	10.94	10.99	10.97	11.29	11.60	11.99	12.37	12.80	13.09	13.56
14.61	14.62	14.62	14.62	14.74	14.74	14.77	14.74	16.00	15.97	16.29	16.60	17.17	17.55	17.98	18.41	18.74
9.31	9.26	9.26	9.26	9.35	9.35	9.35	9.18	9.18	9.18	9.47	9.47	9.56	9.56	9.73	11.07	10.90
16.95	16.91	16.73	16.73	17.02	16.65	16.48	16.65	16.54	17.11	17.02	16.74	17.60	17.97	22.96	23.67	23.69
20.94	20.90	20.72	20.72	21.05	20.68	20.51	20.68	20.89	21.14	21.05	20.77	21.67	22.03	24.02	24.73	24.75
17.88	17.88	17.88	17.77	18.52	18.52	18.52	18.52	18.52	19.05	19.05	19.05	20.37	20.92	22.76	24.20	24.42
17.74	17.74	17.74	17.64	18.45	18.45	18.45	18.45	18.45	18.98	18.98	18.98	20.37	20.92	22.76	24.20	24.42
12.12	12.12	11.90	11.91	11.91	11.91	11.91	11.91	11.91	12.65	12.65	12.65	15.80	19.89	19.56	21.04	21.22
10.09	10.09	10.88	10.88	10.88	10.88	10.88	10.90	12.10	12.23	11.72	13.83	14.15	14.93	14.99	15.13	16.83
11.48	11.48	12.22	12.22	12.22	12.22	12.22	12.24	13.57	13.70	13.19	15.39	15.71	16.49	16.55	16.69	18.48
17.99	16.19	16.19	16.19	16.19	16.19	16.19	16.19	16.30	16.30	16.30	18.00	18.00	18.77	19.98	20.16	20.12
11.60	11.60	11.60	11.60	11.60	11.60	11.60	11.60	11.67	11.67	11.67	12.39	12.39	12.56	13.77	13.95	13.89
10.79	10.79	10.79	10.79	10.79	10.79	10.79	10.79	10.86	10.86	10.86	11.53	11.53	11.26	12.47	12.65	12.61
19.64	19.64	19.64	19.64	19.64	19.64	19.64	19.64	19.64	19.64	19.64	24.06	24.83	26.44	26.86	26.59	26.66
22.06	22.06	22.06	22.06	22.06	22.06	22.06	22.06	22.06	22.06	22.06	26.48	27.25	29.47	29.89	29.57	29.69
(26.40	26.40	26.40	26.40	26.40	26.40	26.40	26.40	26.40	26.40	26.40	26.40	27.13	27.13	34.03	34.21	34.39
(27.43	27.43	27.43	27.43	27.43	27.43	27.43	27.43	27.43	27.43	27.43	27.43	28.16	28.16	35.47	35.65	35.84
21.94	21.94	21.94	21.94	21.94	21.94	21.94	21.94	21.94	21.94	21.94	21.94	22.67	22.67	2/		
21.85	21.85	21.85	21.85	21.85	21.85	21.85	21.85	21.85	21.85	21.85	21.85	21.85	21.85	21.85	25.01	25.05
26.39	26.39	26.39	26.39	25.68	25.68	25.68	25.68	25.68	25.68	25.68	25.68	25.68	25.68	25.68	28.85	28.85
30.75	30.75	30.75	30.75	30.75	30.75	30.75	30.75	30.75	30.75	30.75	31.58	31.58	31.58	31.58	36.81	36.97
25.75	25.75	25.75	25.60	25.60	25.60	25.60	25.60	25.60	25.60	25.60	25.66	25.66	25.66	26.43	30.62	30.78
21.94	21.94	21.94	18.90	18.90	18.90	18.90	18.90	18.90	18.90	18.90	22.11	22.11	22.11	22.77	22.57	22.74
25.63	23.35	23.35	23.35	23.35	23.35	22.38	22.41	22.41	22.63	22.63	23.57	23.57	26.93	26.93	27.20	31.31
29.31	27.06	27.06	27.06	27.06	27.06	26.09	26.12	26.12	26.38	26.38	27.32	27.32	30.68	30.68	30.68	35.06
32.91	32.91	32.91	32.91	(30.83	30.83	31.13	31.13	31.13	31.13	31.13	31.13	31.13	31.13	31.43	31.43	40.94
				(27.43	27.43	27.70	27.70	27.70	27.70	27.70	27.70	27.70	27.70	27.96	27.96	40.31
64.90	64.90	64.90	64.90	64.90	66.84	67.50	67.50	66.62	67.72	67.72	67.61	67.61	71.42	39.36 3/	39.36	39.36
71.28	71.28	71.28	71.28	71.28	73.42	74.13	74.13	73.25	74.35	74.35	74.25	74.25	78.05	47.50 3/	47.50	47.50
24.49	24.49	24.49	24.49	24.49	25.22	25.47	25.47	23.93	25.02	25.02	24.91	24.91	24.71	24.71	24.71	24.71
32.63	32.63	32.63	32.63	32.63	32.63	32.95	32.95	33.94	33.49	33.49	33.36	33.36	37.18	32.70	32.70	32.70
33.65	33.11	33.11	32.95	32.95	31.64	31.95	31.64	31.64	31.68	31.68	31.77	31.77	32.33	33.86	35.10	36.20
33.36	32.84	32.84	32.66	32.66	26.31	26.56	30.43	30.43	30.46	30.46	30.56	30.56	31.10	32.56	33.76	34.82

Table 2
Sheet 2 of 4

TYPICAL MONTHLY BILLS PAID BY RESIDENTIAL CUSTOMERS FOR CENTRAL AIR CONDITIONING
GAS AND ELECTRICITY
($/Month)

Census Region SMSA	1960 Jan.	1960 July	1961 Jan.	1961 July	1962 Jan.	1962 July	1963 Jan.	1963 July	1964 Jan.	1964 July
East North Central										
Chicago, Ill.										
Gas[5]										
with gas sub. base load	17.23	17.22	17.22	17.20	17.57	16.86	17.17	16.59	16.83	12.45
with elec. sub. base load	18.98	18.97	18.97	18.94	19.31	18.61	18.91	18.33	18.58	14.23
Electric										
without elec. space heating	22.52	22.52	22.52	22.52	22.52	22.52	22.52	22.52	22.52	20.03
with elec. space heating	16.15	16.15	16.15	16.15	16.15	16.15	16.04	16.04	16.04	12.56
Detroit, Mich.										
Gas										
without gas space heating										
with gas sub. base load	17.15	17.15	17.99	17.99	17.99	17.99	17.43	17.43	17.43	17.43
with elec. sub. base load	18.27	18.27	19.36	19.36	19.36	19.36	18.82	18.82	18.82	18.82
with gas space heating										
with gas sub. base load	14.80	14.80	15.47	15.47	15.47	15.47	15.47	15.47	15.47	15.47
with elec. sub. base load	16.18	16.18	17.17	17.17	17.17	17.17	17.08	17.08	17.08	17.08
Electric										
without elec. space heating										
with gas sub. base load	22.76	22.76	22.98	22.98	22.98	22.98	22.98	22.98	22.98	22.98
with elec. sub. base load	21.86	21.86	22.07	22.07	22.07	22.07	22.07	22.07	22.07	22.07
with elec. space heating									17.66[6]	17.66
Columbus, Ohio										
Gas										
with gas sub. base load	18.70	18.95	19.44	19.56	19.68	20.66	19.81	19.07	19.19	20.55
with elec. sub. base load	19.44	19.69	20.17	20.29	20.41	21.39	20.52	19.78	19.90	21.26
Electric										
without elec. space heating										
with gas sub. base load with elec. sub. base load	28.24	28.24	28.24	28.24	28.24	28.24	28.24	28.24	28.24	28.24
with elec. space heating										
with gas sub. base load	26.41	26.41	26.41	26.41	26.41	26.41	26.41	26.41	24.93	24.93
with elec. sub base load	22.29	22.29	22.29	22.29	22.29	22.29	22.29	22.29	20.81	20.81
East South Central										
Nashville, Tenn.										
Gas										
with gas sub. base load	14.77	14.77	14.77	14.77	14.77	14.77	14.77	14.77	14.77	14.77
with elec. sub. base load	16.93	17.09	17.09	17.09	17.09	17.09	17.09	16.93	16.93	16.93
Electric										
with gas sub. base load	10.34	10.34	10.34	10.34	10.34	10.34	10.34	10.34	10.34	10.34
with elec. sub. base load	13.23	13.23	13.23	13.23	13.23	13.23	13.23	13.23	13.23	13.23
West South Central										
Houston, Tex.										
Gas										
with gas sub. base load	25.53	16.29	16.29	16.29	16.29	16.29	16.29	16.29	16.29	16.29
with elec. sub. base load	27.83	19.98	19.98	19.98	19.98	19.98	19.98	19.98	19.98	19.98
Electric	37.70	38.96	38.96	39.41	40.20	40.70	40.70	37.65	37.65	37.65
New Orleans										
Gas										
with gas sub. base load	18.31	19.40	19.40	18.89	21.77	21.77	21.77	21.77	21.77	21.77
with elec. sub. base load.	18.68	20.91	20.91	21.40	21.40	21.40	21.40	21.40	21.40	23.28
Electric										
with gas sub. base load	41.32	41.32	43.43	43.43	43.82	43.82	44.35	44.35	43.17	43.17
with elec. sub. base load	40.81	40.81	42.92	42.92	43.31	43.31	43.84	43.84	42.66	42.66
Tulsa										
Gas										
with gas sub. base load	22.15	22.40	22.28	20.28	20.28	20.32	20.32	20.48	20.60	20.69
with elec. sub. base load	23.98	24.23	24.10	21.88	21.88	21.92	21.92	22.08	22.21	22.29
Electric										
without elec. space heating										
with gas sub. base load with elec. sub. base load	59.74	59.74	59.74	59.87	59.87	60.45	60.45	60.45	60.28	(49.54 (48.69
with elec. space heating	24.07	24.07	24.07	24.20	24.20	24.78	24.78	24.78	24.61	24.61
West North Central										
Minneapolis-St. Paul, Minn.										
Gas										
with gas sub. base load	13.54	13.54	13.54	13.54	13.54	13.54	13.54	13.54	13.54	13.54
with elec. sub. base load	15.66	15.66	15.66	15.66	15.66	15.66	15.66	15.66	15.66	15.66
Electric										
with gas sub. base load	11.43	11.43	11.43	11.43	11.43	11.43	11.43	11.43	11.43	11.43
with elec. sub. base load	12.58	12.58	12.58	12.58	12.58	12.58	12.58	12.58	12.58	12.58

1965		1966		1967		1968		1969		1970		1971		1972		1973
Jan.	July	Jan.	July	Jan.	July	Jan.	July	Jan.	July	Jan.	July	Jan.	July	Jan.	July	Jan.
12.43	12.36	12.49	12.28	12.44	12.44	12.84	12.94	11.82	11.48	13.51	13.08	14.06	14.09	15.48	12.73	13.64
14.21	14.14	14.29	14.09	14.25	14.25	14.67	14.76	13.64	13.30	15.34	14.90	15.89	15.91	17.30	14.48	15.34
20.03	20.03	19.34	19.26	19.37	19.06	19.68	19.68	19.68	19.68	19.68	19.58	21.19	22.30	24.71	24.99	24.88
12.56	12.56	12.29	12.20	12.30	12.07	11.51	11.51	11.51	11.51	11.51	11.42	11.51	12.60	13.60	13.88	13.77
17.43	17.43	17.43	17.43	17.43	17.43	17.43	17.43	17.43	17.43	17.43	17.43	17.43	17.32	17.32	17.32	19.01
18.82	18.82	18.82	18.82	18.82	18.82	18.82	18.82	18.82	18.82	18.82	18.82	18.82	19.09	19.09	19.09	20.97
15.34	15.34	15.25	15.25	15.25	15.25	15.25	15.25	15.25	15.25	15.25	15.25	15.25	15.32	15.32	15.32	17.19
16.96	16.96	16.88	16.88	16.88	16.88	16.88	16.88	16.88	16.88	16.88	16.88	16.88	17.34	17.34	17.34	19.34
22.98	18.34	19.95	19.95	19.95	19.42	19.42	19.42	19.42	19.42	19.95	19.95	19.95	21.17	21.17	21.17	21.17
22.07	17.66	16.56	16.56	16.56	16.56	16.56	16.56	16.56	16.56	17.66	17.66	17.66	20.97	20.97	20.97	20.97
17.66	16.56	16.56	16.56	16.56	14.89	14.89	14.89	14.89	14.89	15.44	15.44	15.44	16.56	16.56	16.56	16.56
19.30	15.62	17.23	17.96	18.31	18.46	18.70	17.75	16.93	18.23	18.75	19.85	20.17	21.65	21.93	24.37	24.57
20.01	16.33	17.95	18.67	19.03	19.17	19.42	18.47	17.65	18.95	19.46	20.57	20.78	22.36	22.64	25.08	25.29
28.24	28.24	(26.09	26.09	26.22	26.22	26.34	26.34	26.34	26.34	26.34	26.34	26.34	28.60	30.50	30.50	31.32
		(25.27	25.27	25.39	25.39	25.51	25.51	25.51	25.51	25.51	25.51	25.51	27.77	29.67	29.67	30.49
24.93	24.93	23.61	23.61	23.72	23.72	23.84	23.84	23.84	23.84	23.84	23.84	23.84	23.84	25.74	25.74	26.56
20.81	20.81	17.84	17.84	17.93	17.93	18.01	18.01	18.01	18.01	18.01	18.01	18.01	18.01	19.92	19.92	20.74
14.77	14.77	14.77	14.77	14.77	14.77	14.77	14.77	14.77	14.77	14.77	14.77	14.77	19.13	19.13	19.13	17.97
16.93	16.93	16.93	16.93	16.93	16.93	16.93	16.93	16.93	16.93	16.93	16.93	16.93	21.30	21.30	21.30	20.56
10.34	10.34	10.34	10.34	10.34	10.34	11.51	11.51	11.51	13.72	13.72	13.72	17.48	17.48	17.48	17.48	17.48
13.23	13.23	13.23	13.23	13.23	13.23	13.43	13.43	13.43	13.41	13.41	13.41	17.11	17.11	17.11	17.11	17.11
16.29	16.29	16.29	16.29	16.29	16.29	16.45	16.45	16.91	16.91	16.91	16.91	16.99	16.99	20.17	20.17	20.17
19.98	19.98	19.98	19.98	19.98	19.98	20.18	20.18	21.57	21.57	21.57	21.57	21.67	21.67	25.02	25.02	25.02
37.65	37.38	37.38	37.59	37.59	37.59	37.96	37.96	38.51	38.71	38.71	38.71	38.89	38.89	38.96	38.96	40.13
21.77	21.77	21.77	21.77	21.77	21.77	22.41	22.41	22.41	22.41	22.62	22.62	22.62	28.42	28.42	31.51	31.61
23.28	23.28	23.28	23.28	23.28	23.28	23.96	23.96	23.96	23.96	24.19	24.19	24.19	29.99	29.99	33.08	33.18
42.77	42.77	42.11	42.11	41.98	41.98	43.22	43.22	43.22	43.22	43.63	42.94	42.94	42.94	43.76	44.58	44.58
42.26	42.26	41.60	41.60	42.79	42.79	44.05	44.05	44.05	44.05	44.67	42.41	42.41	42.41	43.23	44.05	44.05
20.93	21.01	20.93	21.05	21.05	21.18	21.22	21.42	22.11	22.11	22.03	22.03	22.15	22.39	22.39	23.89	24.10
23.18	23.26	21.12	23.30	23.30	23.42	23.48	23.69	24.77	24.77	24.69	24.69	24.81	25.09	25.05	26.55	26.76
50.03	50.03	49.52	49.52	49.41	49.41	49.65	49.65	49.75	49.62	49.62	49.62	49.75	49.75	51.58	51.58	52.44
49.17	49.17	46.36	46.01	48.14	48.14	48.48	48.48	48.55	48.45	48.45	48.45	48.59	48.59	50.41	50.41	51.26
24.61	24.61	48.36	46.01	45.79	45.79	46.13	46.13	46.21	46.10	46.10	46.10	46.24	46.24	48.04	48.04	48.89
13.11	13.11	12.41	12.78	12.78	12.78	12.78	12.78	12.78	13.36	13.36	13.36	13.36	14.38	14.52	14.52	14.52
15.66	15.31	15.31	14.81	15.25	15.25	15.25	15.25	15.25	16.40	16.40	16.40	16.40	17.95	18.13	18.13	18.13
11.43	11.43	11.43	11.43	11.77	11.77	11.77	11.77	11.77	11.77	11.77	11.77	11.77	11.77	14.55	14.65	14.86
12.58	12.58	12.58	12.58	12.96	12.96	12.96	12.96	12.96	12.96	12.96	12.96	12.96	12.96	13.23	13.34	13.54

Table 2
Sheet 3 of 4

TYPICAL MONTHLY BILLS PAID BY RESIDENTIAL CUSTOMERS FOR CENTRAL AIR CONDITIONING
GAS AND ELECTRICITY
($/Month)

Census Region SMSA	1960 Jan.	1960 July	1961 Jan.	1961 July	1962 Jan.	1962 July	1963 Jan.	1963 July	1964 Jan.	1964 July
West North Central (Continued)										
Omaha, Neb.										
Gas										
without gas space heating										
with gas sub. base load	20.05	20.05	20.05	20.05	20.05	20.05	20.05	18.54	18.54	18.54
with elec. sub. base load	22.09	22.09	22.09	22.09	22.09	22.09	22.09	20.56	20.56	20.56
with gas space heating										
with gas sub. base load	15.15	15.15	15.15	15.15	15.15	15.15	15.15	14.54	14.54	14.54
with elec. sub. base load	17.75	17.75	17.75	17.75	17.75	17.75	17.75	17.00	17.00	17.00
Electric										
Publicly owned	26.63	26.63	26.63	26.63	26.63	26.63	26.63	26.63	26.63	26.63
Privately owned										
with gas sub. base load	31.51	31.51	31.51	31.24	31.24	31.33	25.43	25.43	25.15	25.15
with elec. sub. base load	31.32	31.32	31.32	31.06	31.06	31.15	23.82	23.82	23.54	23.54
St. Louis										
Gas										
without gas space heating										
with gas sub. base load	43.87	44.20	44.20	44.20	44.63	44.63	44.28	44.68	44.68	45.12
with elec. sub. base load	45.44	45.99	45.99	45.99	46.44	46.44	46.08	46.48	46.48	46.94
with gas space heating										
with gas sub. base load	28.13	29.22	29.22	29.22	29.51	29.51	29.16	29.56	29.56	29.85
with elec. sub. base load	31.21	32.48	32.48	32.48	32.80	32.80	32.45	32.85	32.85	33.17
Electric	45.24	45.24	45.24	45.24	45.69	45.69	45.69	45.69	45.69	46.13
Mountain										
Denver										
Gas										
with gas sub. base load	5.71	5.19	6.46	6.46	6.46	6.46	6.46	6.46	6.46	6.46
with elec. sub. base load	6.65	6.13	8.14	8.14	8.14	8.14	8.14	8.14	8.14	8.14
Electric										
without elec. space heating										
with gas sub. base load				(17.70	17.70	17.70	17.70	17.70	17.70	17.70
with elec. sub. base load	16.85	16.85	16.85	(						
with elec. space heating				(12.64	12.64	12.64	12.64	12.64	12.64	12.64
Phoenix, Ariz.										
Gas										
with gas sub. base load	25.90	25.90	35.44	35.44	35.18	35.18	34.46	34.46	33.62	34.93
with elec. sub. base load	28.84	28.84	38.37	38.37	35.71	35.71	37.38	37.38	36.54	37.73
Electric										
with gas sub. base load / with elec. sub. base load	52.57	52.57	53.03	53.03	53.03	48.56	48.56	53.02	53.02	53.02
Salt Lake City, Utah										
Gas										
with gas sub. base load	10.20	10.64	10.69	10.69	10.70	10.70	10.75	10.75	10.75	10.65
with elec. sub. base load	12.13	12.65	12.71	12.71	12.72	12.72	12.78	12.78	12.78	12.65
Electric										
with gas sub. base load	17.97	17.97	18.06	18.06	18.06	18.06	18.14	23.01	23.01	23.01
with elec. sub. base load	21.17	21.17	21.27	21.27	21.27	21.27	21.37	19.30	19.30	19.30
Pacific										
Los Angeles, Calif.										
Gas										
with gas sub. base load	5.30	5.30	5.30	5.30	5.30	5.30	5.30	4.61	4.61	4.45
with elec. sub. base load	7.50	7.50	7.50	7.50	7.50	7.50	7.50	6.94	6.94	6.77
Electric										
publicly owned										
with gas sub. base load	5.53	5.53	5.53	5.53	5.70	5.70	5.70	5.70	5.65	5.43
with elec. sub. base load	4.26	4.26	4.26	4.26	4.42	4.42	4.42	4.42	4.39	4.16
privately owned										
with gas sub. base load	5.91	5.91	5.91	5.91	5.91	5.91	5.91	5.91	5.91	5.91
with elec. sub. base load	4.65	4.65	4.65	4.65	4.65	4.65	4.65	4.65	4.65	4.65
San Francisco, Calif.										
Gas										
with gas sub. base load	1.33	1.33	1.33	1.33	1.33	1.33	1.33	1.39	1.38	1.38
with elec. sub. base load	2.18	2.18	2.18	2.18	2.18	2.18	2.18	2.44	2.36	2.36
Electric										
with gas sub. base load / with elec. sub. base load	1.55	1.55	1.55	1.55	1.55	1.55	1.55	1.55	1.55	2.11
Seattle, Wash.										
Gas										
without gas space heating										
with gas sub. base load	7.16	7.16	7.16	7.16	7.16	6.74	6.74	6.74	6.74	6.74
with elec. sub. base load	11.94	11.94	11.94	11.94	11.94	11.52	11.52	11.52	11.52	11.52
with gas space heating										
with gas sub. base load	5.37	5.37	5.37	5.08	5.08	5.08	5.08	4.90	4.90	4.90
with elec. sub. base load	9.58	9.58	9.58	9.58	9.58	9.58	9.58	6.79	6.79	6.79

1965		1966		1967		1968		1969		1970		1971		1972		1973
Jan.	July	Jan.	July	Jan.	July	Jan.	July	Jan.	July	Jan.	July	Jan.	July	Jan.	July	Jan.
16.50	16.50	18.54	18.54	18.54	16.89	17.31	17.31	17.50	18.32	18.32	18.78	18.78	20.99	20.99	20.99	20.99
18.30	18.30	20.56	20.56	20.56	18.75	19.22	19.22	19.43	20.58	20.58	21.03	21.03	23.33	23.33	23.33	23.33
12.94	12.94	14.54	14.54	14.54	14.54	14.90	14.90	15.22	15.22	15.22	15.53	15.33	18.94	18.94	18.94	18.94
15.13	15.13	17.00	17.00	17.00	16.68	17.10	17.10	17.43	17.84	17.84	18.15	18.15	21.50	21.50	21.50	21.50
26.63	26.63	26.63	26.63	26.63	27.30	27.30	27.30	27.30	27.30	27.30	27.30	29.29	29.29	29.29	29.29	29.29
25.15	25.15	25.15	25.43	25.43	25.43	25.68	25.68	25.43	25.40	25.42	25.42	25.42	31.94	35.33	36.05	36.29
23.54	23.54	23.54	23.82	23.82	23.82	24.05	24.05	23.77	23.77	23.77	23.77	23.77	28.85	34.04	34.76	35.00
46.02	46.00	46.00	45.94	45.94	45.94	45.94	45.94	46.82	46.82	49.31	51.87	51.87	55.11	56.10	56.10	59.20
47.91	47.90	47.90	47.84	47.84	47.84	47.84	47.84	48.76	48.76	51.43	54.00	54.00	57.23	58.28	58.27	61.35
30.14	30.14	30.14	30.08	30.08	30.08	30.08	30.08	30.66	30.66	32.82	35.38	35.38	38.63	39.45	39.49	42.56
33.57	33.57	33.57	33.51	33.51	33.51	33.51	33.51	34.16	34.16	36.55	39.11	39.11	42.35	43.22	43.25	46.32
46.13	46.13	46.13	46.13	46.13	46.13	44.83	44.83	44.83	44.83	47.43	47.43	47.43	47.43	49.53	49.53	49.53
6.52	6.52	6.52	6.52	6.71	6.71	6.71	6.71	6.71	6.71	6.75	6.75	6.71	6.71	7.43	7.43	7.43
8.18	8.18	8.18	8.18	8.42	8.42	8.42	8.42	8.42	8.42	8.47	8.47	8.42	8.42	9.48	9.48	9.48
17.87	17.87	17.87	17.87	18.39	18.39	(18.14	18.14	18.14	18.14	18.14	18.14	18.14	18.14	18.14	20.31	20.31
						(10.50	10.50	10.50	10.50	10.50	10.50	10.50	10.50	10.50	13.13	13.13
12.76	12.76	12.76	12.76	13.13	13.13	(10.50	10.50	10.50	10.50	10.50	10.50	10.50	10.50	10.50	13.13	13.13
33.91	33.91	34.65	34.65	34.65	34.55	33.87	34.27	34.77	33.95	35.78	35.78	35.78	39.35	32.90	32.90	32.90
36.84	36.84	37.64	37.64	37.64	37.54	36.90	37.30	37.85	36.98	38.81	38.81	38.81	45.46	37.05	37.05	37.05
53.24	53.24	53.13	53.13	53.45	53.45	54.51	54.51	(53.42	53.54	53.54	53.54	53.83	53.83	54.37	55.18	55.18
								(52.17	52.31	52.31	52.31	52.59	52.59	53.14	53.93	53.93
10.65	10.65	10.65	10.65	10.65	10.65	10.65	10.65	9.21	9.21	9.03	9.21	9.21	9.21	9.21	10.13	10.13
12.65	12.65	12.65	12.65	12.65	12.65	12.65	12.65	9.62	9.62	9.44	9.63	9.63	9.63	9.63	10.54	10.54
23.01	23.01	23.01	23.01	23.01	23.01	23.01	23.01	23.23	23.23	23.23	23.23	23.23	23.23	23.23	27.11	27.11
19.30	19.30	19.30	19.30	19.30	19.30	19.30	19.30	15.31	15.31	15.31	15.31	15.31	15.31	15.31	17.37	17.37
4.45	4.45	4.31	4.31	4.19	4.19	4.19	4.19	4.19	4.32	4.31	4.31	4.84	4.84	4.84	4.84	5.31
6.77	6.77	6.65	6.65	6.45	6.45	6.45	6.45	6.45	6.57	6.56	6.56	7.70	7.70	7.70	7.70	8.37
5.43	5.43	5.43	5.43	5.74	5.74	5.74	5.74	5.77	5.77	5.84	5.84	5.84	5.84	6.19	6.56	6.60
4.16	4.16	4.16	4.16	4.40	4.40	4.40	4.40	4.43	4.43	4.50	5.19	5.19	5.18	5.52	5.89	5.94
5.91	5.91	5.91	5.91	5.91	5.91	6.03	6.03	6.03	6.03	6.17	6.17	6.17	6.17	6.92	7.06	7.35
4.65	4.65	4.65	4.65	4.65	4.65	4.74	4.74	4.74	4.74	5.30	5.30	5.30	5.30	5.62	5.76	6.08
1.38	1.38	1.38	1.38	1.32	1.32	1.37	1.37	1.37	1.14	1.14	1.14	1.14	1.14	1.14	1.14	1.14
2.36	2.36	2.43	2.43	2.42	2.42	2.51	2.51	2.51	2.30	2.40	2.40	2.40	2.40	2.39	2.39	2.39
2.11	2.11	2.11	2.11	2.15	2.15	2.15	2.15	2.15	2.15	(2.23	2.23	2.23	1.92	1.95	1.95	1.95
										(2.05	2.05	2.05	1.66	1.69	1.69	1.69
6.75	6.75	6.81	6.81	6.81	6.81	6.83	6.83	6.90	6.90	6.90	6.90	6.90	6.90	6.90	6.90	5.68
11.65	11.65	11.65	11.65	11.65	11.65	11.68	11.68	11.80	11.80	11.80	10.83	10.83	10.83	10.83	10.83	8.57
4.76	4.76	4.80	4.50	4.50	4.37	4.38	4.38	4.42	4.42	4.42	4.42	4.42	4.42	4.42	4.42	5.01
5.70	5.70	5.75	5.75	5.75	5.75	5.77	5.77	5.82	5.82	5.82	5.56	5.56	5.56	5.56	5.56	5.01

Table 2 Sheet 4 of 4
TYPICAL MONTHLY BILLS PAID BY RESIDENTIAL CUSTOMERS FOR CENTRAL AIR CONDITIONING GAS AND ELECTRICITY (FOOTNOTES)

1/ New rate available.
2/ All electric home rate no longer offered.
3/ A new residential rate became available to this class of customers.
4/ Special promotional rates not available.
5/ Spring cooling concessional rates 7/1/64-1/1/73.
6/ All electric home rate became available.
Source: American Gas Association Rate Service. Federal Power Commission National Electric Rate Book.

Table 3
RETAIL SERVICE STATION PRICES* OF MAJOR BRAND REGULAR GRADE GASOLINE
(¢/Gal.)

Census Region SMSA	1960 Jan.	1960 July	1961 Jan.	1961 July	1962 Jan.	1962 July	1963 Jan.	1963 July	1964 Jan.	1964 July	1965 Jan.	1965 July
New England												
Boston, Mass.	22.9	27.9	27.9	28.9	27.9	25.9	25.9	28.9	28.9	28.9	26.9	29.9
Middle Atlantic												
Albany, N.Y.	29.9	29.9	28.9	28.9	26.9	29.9	28.9	26.9	27.9	28.9	29.9	29.9
New York, N.Y.	33.9	31.9	31.9	31.9	31.9	31.9	31.9	31.9	31.9	31.9	31.9	31.9
Pittsburgh, Pa.	30.9	30.2	30.4	30.6	29.9	29.9	28.9	30.9	29.9	26.9	27.9	30.9
South Atlantic												
Atlanta, Ga.	33.9	30.9	31.9	31.9	31.9	30.9	33.9	32.9	32.9	32.9	32.9	32.9
Charlotte, N.C.	31.9	30.9	31.9	31.9	31.9	30.9	29.9	31.9	31.9	29.9	31.9	31.9
Washington, D.C.	29.9	29.9	30.9	29.9	30.9	30.9	29.9	30.9	29.9	29.9	29.9	30.9
Miami, Fla.	32.9	31.9	31.9	31.9	31.9	31.9	32.9	32.9	32.9	32.9	32.9	32.9
East North Central												
Chicago, Ill.	31.9	32.9	32.9	32.9	31.9	28.9	32.9	32.9	29.9	30.9	31.9	33.9
Detroit, Mich.	27.9	33.9	33.9	26.9	24.9	25.9	29.9	33.9	28.9	28.9	28.9	33.9
Columbus, Ohio	31.9	31.9	32.9	32.9	31.9	31.9	31.9	31.9	31.9	31.9	31.9	31.9
East South Central												
Nashville, Tenn.	30.9e/	30.9e/	30.9e/	30.9e/	30.9e/	30.9e/	30.9e/	30.9e/	31.9e/	31.9e/	31.9e/	31.9e/
West North Central												
Minneapolis-St. Paul, Minn.	29.9	31.4	31.4	29.4	29.4	26.4	29.4	31.4	29.4	28.4	28.4	29.4
Omaha, Neb.	31.9	31.9	25.9	33.9	33.9	32.9	31.9	34.9	31.9	31.9	27.9	33.9
St. Louis, Mo.	24.9	29.9	30.9	22.9	30.9	19.9	30.9	30.9	24.9	27.9	29.9	29.9
West South Central												
Houston, Tex.	29.9	29.9	30.9	29.9	28.9	28.9	27.9	27.9	27.9	27.9	26.9	28.9
New Orleans, La.	29.9	26.9	29.9	26.9	29.9	29.9	29.9	29.9	29.9	29.9	29.9	29.9
Tulsa, Okla.	31.9	32.9	32.9	31.9	31.9	32.9	29.9	32.9	31.9	29.9	31.9	31.9
Mountain												
Denver, Colo.	27.9	33.9	33.9	32.9	29.9	27.9	32.9	31.9	25.9	32.9	29.9	31.9
Phoenix, Ariz.	25.9	26.9	22.9	28.9	24.9	32.9	31.9	29.9	26.9	26.9	33.9	32.9
Salt Lake City, Utah	31.9	33.9	33.9	32.9	32.9	31.9	32.9	31.9	32.9	32.9	32.9	32.9
Pacific												
Los Angeles, Calif.	29.9	29.9	29.9	29.9	29.9	30.9	30.9	25.9	32.9	30.9	31.9	33.9
San Francisco, Calif.	32.9	31.9	29.9	27.9	31.9	31.9	31.9	31.9	32.9	32.9	32.9	33.9
Seattle, Wash.	29.9	30.9	32.9	26.9	29.9	32.9	31.9	31.9	31.9	31.9	28.9	32.9

e/ Estimated.

*Includes taxes.

Note: Prices are representative as of first of month.

Source: Platt's Oilgram Price Service; Oil and Gas Journal; Commerce Clearing House; various oil companies.

1966		1967		1968		1969		1970		1971		1972		1973
Jan.	July	Jan.	July	Jan.	July	Jan.	July	Jan.	July	Jan.	July	Jan.	July	Jan.
29.9	30.9	31.9	32.4	32.9	32.9	33.9	34.9	34.9	35.9	36.9	36.9	37.9	34.9	35.9
30.9	31.9	32.9	33.9	33.9	33.9	34.9	35.9	37.7	34.9	35.9	35.9	34.9	38.9	38.9
32.9	32.9	32.9	34.5	34.9	35.9	36.9	37.9	37.9	38.9	39.9	39.9	41.9	41.9	41.9
30.9	31.9	31.9	31.9	32.9	32.9	33.9	34.9	34.9	36.9	37.9	35.9	37.8	37.9	37.9
32.9	33.9	33.9	33.9	33.9	33.9	35.9	35.9	35.9	35.9	36.9	37.9	38.9	36.9	38.9
31.9	32.9	32.9	33.9	33.9	33.9	33.9	36.9	35.9	37.9	38.9	34.9	38.9	36.9	38.9
30.9	31.9	32.9	33.3	33.2	33.9	34.9	34.9	34.9	35.9	36.9	36.9	37.9	37.9	38.9
32.9	33.9	29.9	27.9	33.9	33.9	29.9	34.9	34.9	35.9	36.9	35.9	37.9	35.9	32.9
33.9	33.9	33.9	34.9	35.9	35.9	35.9	36.9	38.4	39.4	40.4	38.4	37.3	33.1	40.4
33.9	33.9	33.9	34.9	36.9	35.9	35.9	32.9	32.9	38.9	38.9	31.9	31.9	36.9	36.9
31.9	32.9	32.9	33.9	33.9	34.9	34.9	35.9	35.9	36.9	31.9	37.9	37.9	35.9	37.9
31.9[e]	33.9[e]	33.9	33.9	33.9	34.9	35.9	35.9	36.9	36.9	37.9	31.9	35.9	35.9	35.9
31.9	32.9	33.9	33.9	35.9	34.9	34.9	36.9	34.9	37.9	35.9	36.9	38.9	36.9	36.9
33.9	34.9	34.9	35.9	35.9	33.9	35.9	36.9	37.9	38.9	34.9	39.9	30.9	37.9	37.9
30.9	32.9	32.9	33.9	33.9	33.9	33.9	34.9	34.9	35.9	36.9	36.9	36.9	34.9	38.9
28.9	29.9	29.9	29.9	29.9	30.9	30.9	26.9	31.9	32.9	33.9	33.9	33.9	31.9	31.9
29.9	30.9	31.9	31.9	31.9	32.9	32.9	34.9	34.9	35.9	36.9	36.9	32.9	34.9	34.9
31.9	31.9	32.9	33.9	33.9	33.9	33.9	34.9	34.9	35.9	36.9	36.9	36.9	36.9	36.9
29.9	32.9	34.9	34.9	34.9	31.9	34.9	32.9	28.9	37.9	36.9	35.9	35.9	36.9	34.9
31.9	33.9	34.9	34.9	34.9	34.9	34.9	35.9	35.9	36.9	37.9	37.9	37.9	35.9	36.9
32.9	29.9	27.9	32.9	32.9	32.9	32.9	34.9	34.9	35.9	36.9	36.9	36.9	36.9	36.9
28.9	32.9	30.9	33.9	33.9	33.9	33.9	35.9	34.9	35.9	36.9	36.9	34.9	34.9	37.9
32.9	32.9	34.9	34.9	34.9	34.9	34.9	36.9	35.9	35.9	36.9	36.9	35.9	36.9	38.9
30.9	28.9	32.9	33.9	33.9	35.9	35.9	36.9	36.9	37.9	38.9	38.9	36.9	34.9	36.9

Table 4-a
Sheet 1 of 2

MONTHLY FUEL BILLS OF SMALL COMMERCIAL CUSTOMERS
GAS AND OIL

($/Month)

Census Region SMSA	1960		1961		1962		1963		1964		1965	
	Jan.	July	Jan.	July	Jan.	July	Jan.	July	Jan.	July	Jan.	July
New England												
Boston, Mass.												
Gas	153.25	153.25	156.26	156.26	149.75	149.75	149.75	149.75	155.46	152.97	152.76	152.76
Oil	77.90		81.60		80.00		82.20		81.60		84.30	
Middle Atlantic												
Albany, N.Y.												
Gas	82.10	82.10	85.31	85.31	85.31	86.14	83.79	82.95	84.05	83.49	84.39	84.45
Oil	72.60		75.80		77.40		80.60		82.20		83.20	
New York, N.Y.												
Gas	117.85	117.85	116.33	117.48	118.49	117.55	118.91	115.49	115.56	119.66	119.66	119.44
Oil	77.90		80.60		80.60		82.70		80.60		82.20	
Pittsburgh, Pa.												
Gas	46.84	46.84	46.84	46.34	46.34	48.65	48.44	48.30	48.23	46.15	46.11	48.20
Oil	--		--		--		--		--		--	
South Atlantic												
Atlanta, Ga.												
Gas	64.89	64.89	64.89	64.89	64.89	64.89	64.89	64.89	64.89	64.89	64.89	64.89
Oil	--		--		--		--		--		--	
Charlotte, N.C.												
Gas	92.94	92.94	92.94	95.41	95.41	95.41	90.78	90.78	88.52	88.52	88.52	88.52
Oil	75.80		79.00		77.40		80.60		74.20		74.20	
Washington, D.C.												
Gas	98.15	98.15	98.15	98.15	99.12	99.12	99.33	99.33	99.33	99.33	99.33	94.82
Oil	75.80		79.50		80.00		82.20		79.50		79.50	
Miami, Fla.												
Gas	108.49	108.49	108.49	108.49	57.14[1/]	57.14	57.14	57.14	57.14	57.14	57.14	57.14
Oil	--		--		--		--		--		--	
East South Central												
Nashville, Tenn.												
Gas	87.48	97.30	97.30	97.30	97.30	97.30	97.30	87.48	87.48	87.48	87.48	87.48
Oil	--		--		--		--		--		--	
East North Central												
Chicago, Ill.												
Gas	76.70	76.62	76.62	76.55	78.18	75.07	73.61	71.11	72.14	73.30	70.92	70.61
Oil	75.80		77.40		79.00		80.00		79.50		80.60	
Detroit, Mich.												
Gas	81.63	58.63	61.70	61.70	61.70	61.70	61.45	61.45	61.45	61.45	60.95	60.95
Oil	79.50		80.00		79.50		79.50		79.50		80.00	
Columbus, Ohio												
Gas	53.95	54.65	56.03	56.38	56.73	59.53	57.06	54.96	55.31	59.17	55.66	45.15
Oil	--		--		--		--		--		--	
West South Central												
Houston, Tex.												
Gas	44.76	44.76	44.76	44.76	45.66	45.66	47.99	47.99	47.99	47.99	47.99	47.99
Oil	--		--		--		--		--		--	
New Orleans, La.												
Gas	32.25	35.66	35.66	35.92	39.43	39.43	39.43	39.43	39.43	39.43	39.43	39.43
Oil	--		--		--		--		--		--	
Tulsa, Okla.												
Gas	46.01	46.44	46.23	46.44	46.44	46.51	46.51	46.80	47.01	47.15	48.32	48.46
Oil	--		--		--		--		--		--	
West North Central												
Minneapolis-St. Paul Minn.,												
Gas	64.92	64.92	64.92	64.92	64.92	64.92	64.92	64.92	64.92	64.92	64.92	64.10
Oil	77.90		76.90		76.30		76.30		76.90		76.30	
Omaha, Neb.												
Gas	56.50	56.50	56.50	56.50	56.50	56.50	56.50	52.43	52.43	52.43	46.66	46.66
Oil	--		--		--		--		--		--	
St. Louis, Mo.												
Gas	56.87	58.14	58.14	58.14	58.71	58.71	58.05	58.80	58.80	59.38	57.74	57.74
Oil	78.40		79.00		80.00		80.60		81.10		81.10	
Mountain												
Denver, Colo.												
Gas	22.72	25.31	33.24	33.24	33.24	33.24	33.24	33.24	33.24	33.24	33.57	33.57
Oil	--		--		--		--		--		--	

1966		1967		1968		1969		1970		1971		1972		1973
Jan.	July	Jan.	July	Jan.	July	Jan.	July	Jan.	July	Jan.	July	Jan.	July	Jan.
152.76	152.76	151.92	152.33	152.33	152.33	152.33	158.46	158.46	158.84	158.84	180.36	180.36	180.36	193.97
87.50		89.60		91.20		92.80		98.60		103.90		106.50		
84.43	84.47	85.17	85.17	85.35	85.21	86.93	86.77	89.07	91.22	94.47	97.16	100.23	103.33	105.63
84.30		88.50		94.30		95.90		101.20		111.80		113.40		
119.44	118.48	120.35	118.34	117.45	118.34	119.46	120.80	120.35	118.78	123.67	125.63	135.46	139.21	139.28
84.30		87.50		90.10		92.20		97.00		103.90		106.00		
52.83	52.83	52.83	52.83	52.83	52.99	58.62	59.21	64.64	66.81	68.30	71.88	72.19	72.84	80.77
--		--		--		--		--		--		--		--
64.89	64.89	64.89	64.89	67.49	67.49	67.49	67.49	67.49	70.08	71.87	76.57	77.54	76.80	77.10
--		--		--		--		--		--		--		--
88.52	88.52	88.52	88.52	88.52	88.52	88.52	88.52	88.52	88.52	88.52	88.52	88.52	95.90	95.90
75.80		75.80		77.90		82.70		86.40		86.90		86.90		
94.82	94.82	94.82	94.82	93.51	93.58	93.58	93.58	93.58	95.74	95.74	103.43	104.42	96.53	107.09
82.70		86.40		89.00		91.70		97.00		101.20		101.80		
57.14	57.14	57.14	58.85	58.85	56.05	55.60	57.79	57.79	57.57	57.57	65.14	56.88	56.88	56.88
--		--		--		--		--		--		--		--
87.48	87.48	87.48	87.48	87.48	87.48	87.48	87.48	87.48	87.48	87.48	97.39	97.39	104.01	105.14
--		--		--		--		--		--		--		--
64.92	64.73	64.73	65.43	66.43	66.82	62.00	60.53	69.30	69.54	72.50	72.57	78.64	76.60	79.58
81.10		82.70		84.30		85.90		90.10		93.80		95.40		
60.95	60.95	60.95	60.60	60.60	60.60	60.60	60.60	60.60	60.60	60.60	61.56	60.25	61.27	68.62
81.10		84.30		86.40		89.00		92.80		95.90		97.00		
49.76	51.82	52.81	53.23	53.92	51.34	48.92	52.62	54.08	57.22	57.78	62.34	63.10	70.03	70.59
--		--		--		--		--		--		--		--
47.99	47.99	47.99	47.99	48.46	48.46	52.32	52.32	52.32	52.32	52.57	52.57	60.15	60.15	60.15
--		--		--		--		--		--		--		--
39.43	38.92	38.92	38.92	40.07	40.07	40.07	40.07	40.45	40.45	40.45	49.67	49.67	54.59	54.75
--		--		--		--		--		--		--		--
48.32	48.53	48.53	48.75	48.46	48.82	53.65	53.65	53.51	53.51	53.72	54.16	54.68	57.29	57.67
--		--		--		--		--		--		--		--
64.10	60.80	62.62	62.62	62.62	62.62	62.62	66.48	66.48	66.48	66.48	72.07	72.76	72.76	72.76
76.30		80.60		83.20		85.90		89.60		92.80		93.30		
52.43	52.43	48.23	48.23	49.44	49.44	41.42	43.55	43.55	44.63	44.63	50.19	50.19	50.19	50.19
--		--		--		--		--		--		--		--
57.74	57.62	57.62	57.62	57.62	57.62	58.17	58.17	59.09	63.77	63.77	69.44	70.70	70.74	76.12
82.20		84.30		86.90		89.60		93.30		98.60		100.70		
33.57	33.57	34.55	34.55	34.55	34.55	34.55	34.50	34.71	34.71	38.09	38.09	38.44	38.44	38.47
--		--		--		--		--		--		--		--

Table 4-a
Sheet 2 of 2

MONTHLY FUEL BILLS OF SMALL COMMERCIAL CUSTOMERS
GAS AND OIL

($/Month)

Census Region	1960		1961		1962		1963		1964		1965	
SMSA	Jan.	July	Jan.	July	Jan.	July	Jan.	July	Jan.	July	Jan.	July
Mountain (Cont.)												
Phoenix, Ariz.												
Gas	48.56	48.56	52.21	52.21	51.84	51.84	50.99	50.99	49.55	51.51	50.00	50.00
Oil	--		--		--		--		--		--	
Salt Lake City, Utah												
Gas	32.73	34.41	34.58	34.58	34.24	34.24	34.41	34.41	34.41	34.08	34.08	34.08
Oil	--		--		--		--		--		--	
Pacific												
Los Angeles, Calif.												
Gas	60.61	60.61	60.61	60.61	60.61	60.61	60.61	63.23	63.23	61.62	61.62	61.62
Oil	--		--		--		--		--		--	
San Francisco, Calif.												
Gas	45.30	45.30	45.30	45.30	45.30	45.30	45.30	46.90	46.40	46.40	46.40	46.40
Oil	--		--		--		--		--		--	
Seattle, Wash.												
Gas	131.56	131.56	131.56	131.56	131.56	123.35	123.35	123.35	123.35	123.35	123.50	123.50
Oil	89.60		88.50		91.70		93.80		93.80		93.30	

1/ New distributor.

Note: Gas figures based on prices in effect on January 1 and July 1 of the given year.

Source: American Gas Association Rate Book.
Data for oil figures described in Chapter IV.

1966		1967		1968		1969		1970		1971		1972		1973
Jan.	July	Jan.	July	Jan.	July	Jan.	July	Jan.	July	Jan.	July	Jan.	July	Jan.
51.04	51.04	51.04	50.88	48.88	50.54	51.29	50.87	52.78	52.78	52.78	57.95	57.95	64.12	64.12
--		--		--		--		--		--		--		--
34.08	34.08	34.08	34.08	34.08	34.08	40.99	40.99	40.19	40.98	40.98	40.98	40.98	43.62	43.62
--		--		--		--		--		--		--		--
59.49	59.49	62.96	62.96	62.96	62.96	62.96	64.65	64.65	64.65	62.68	62.68	62.53	62.53	66.95
--		--		--		--		--		--		--		--
46.40	46.40	46.20	46.20	46.20	46.20	46.20	47.20	49.00	54.01	54.01	54.01	53.89	53.89	53.89
--		--		--		--		--		--		--		--
124.68	124.68	124.68	116.03	116.36	116.36	117.46	117.46	117.46	117.90	117.90	117.90	117.90	117.90	134.10
94.30		97.00		99.60		103.40		107.60		114.50		115.00		

Table 4-b
PRICES PAID FOR FUEL BY SMALL COMMERCIAL CUSTOMERS
GAS
(¢/Mcf)

Census Region / SMSA	1960		1961		1962		1963		1964		1965	
	Jan.	July	Jan.	July	Jan.	July	Jan.	July	Jan.	July	Jan.	July
New England												
Boston, Mass.	219.1	219.1	223.5	223.5	214.1	214.1	214.1	214.1	222.3	218.7	218.4	218.4
Middle Atlantic												
Albany, N.Y.	117.4	117.4	122.0	122.0	122.0	123.2	119.8	118.6	120.2	119.4	119.4	120.8
New York, N.Y.	168.5	168.5	166.4	168.0	169.4	168.1	170.0	165.2	165.3	171.1	171.1	170.8
Pittsburgh, Pa.	67.0	67.0	67.0	66.3	66.3	69.6	69.3	69.1	69.0	66.0	66.0	68.9
South Atlantic												
Atlanta, Ga.	92.8	92.8	92.8	92.8	92.8	92.8	92.8	92.8	92.8	92.8	92.8	92.8
Charlotte, N.C.	132.9	132.9	132.9	136.4	136.4	136.4	129.8	129.8	126.6	126.6	126.6	126.6
Washington, D.C.	135.9	135.9	135.9	135.9	137.2	137.2	141.9	141.9	141.9	141.9	141.9	135.4
Miami, Fla.	155.1	155.1	155.1	155.1	81.7 1/	81.7	81.7	81.7	81.7	81.7	81.7	81.7
East South Central												
Nashville, Tenn.	125.0	139.0	139.0	139.0	139.0	139.0	139.0	125.0	125.0	125.0	125.0	125.0
East North Central												
Chicago, Ill.	109.6	109.6	109.6	109.5	111.8	107.4	105.3	101.7	103.2	104.8	101.4	101.0
Detroit, Mich.	83.8	83.8	88.1	88.1	88.1	88.1	87.9	87.9	87.9	87.9	87.2	87.2
Columbus, Ohio	77.1	78.1	80.1	80.6	81.1	85.1	81.6	78.6	79.1	84.6	79.6	64.6
West South Central												
Houston, Tex.	64.0	64.0	64.0	64.0	65.3	65.3	68.6	68.6	68.6	68.6	68.6	68.6
New Orleans, La.	47.5	51.0	51.0	51.4	56.4	56.4	56.4	56.4	56.4	56.4	56.4	56.4
Tulsa, Okla.	65.8	66.4	66.1	66.4	66.4	66.5	66.5	66.9	67.2	67.4	69.1	69.3
West North Central												
Minneapolis-St. Paul, Minn.	92.7	92.7	92.7	92.7	92.7	92.7	92.7	92.7	92.7	92.7	92.7	91.7
Omaha, Neb.	80.9	80.9	80.9	80.9	80.9	80.9	80.9	75.0	75.0	75.0	66.7	66.7
St. Louis, Mo.	81.3	83.1	83.1	83.1	84.0	84.0	83.0	84.1	84.1	85.0	82.5	82.5
Mountain												
Denver, Colo.	32.4	36.1	47.5	47.5	47.5	47.5	47.5	47.5	47.5	47.5	48.0	48.0
Phoenix, Ariz.	72.9	72.9	74.1	74.1	74.7	74.7	69.4	69.4	70.7	73.6	71.5	71.5
Salt Lake City, Utah	46.8	49.2	49.5	49.5	49.0	49.0	49.2	49.2	49.2	48.7	48.7	48.7
Pacific												
Los Angeles, Calif.	86.6	86.6	86.6	86.6	86.6	86.6	86.6	90.3	90.3	88.1	88.1	88.1
San Francisco, Calif.	64.4	64.4	64.4	64.4	64.4	64.4	64.4	67.0	66.3	66.3	66.3	66.3
Seattle, Wash.	188.1	188.1	188.1	188.1	188.1	176.4	176.4	176.4	176.4	176.4	176.6	176.6

1/ Change in utility.
2/ Unusually large purchased gas adjustment on July 1, 1971.

Note: Prices are those in effect on January 1 and July 1 of the given year.

Source: AGA *Rate Book*.

1966		1967		1968		1969		1970		1971		1972		1973
Jan.	July	Jan.	July	Jan.	July	Jan.	July	Jan.	July	Jan.	July	Jan.	July	Jan.
218.4	218.4	217.2	217.8	217.8	217.8	217.8	226.6	226.6	227.1	227.1	257.9	257.9	257.9	277.1
120.7	120.8	121.8	121.8	122.1	121.9	124.3	124.1	127.4	130.4	135.1	138.9	143.3	147.8	151.1
170.8	169.4	172.1	169.2	168.0	169.2	170.8	172.7	172.1	169.9	176.8	179.7	193.7	199.1	199.2
75.5	75.5	75.5	75.5	75.5	75.8	83.8	84.7	92.4	95.5	97.7	102.8	103.2	104.2	115.5
92.8	92.8	92.8	92.8	96.5	96.5	96.5	96.5	96.5	100.2	102.8	109.5	110.9	109.8	110.2
126.6	126.6	126.6	126.6	126.6	126.6	126.6	126.6	126.6	126.6	126.6	126.6	126.6	137.1	137.1
135.4	135.4	135.4	135.4	133.5	133.6	133.6	133.6	133.6	136.8	136.8	147.8	149.2	137.9	153.0
81.7	81.7	81.7	84.1	84.I	80.2	79.5	82.6	82.6	82.3	82.3	93.1 2/	81.3	81.3	81.3
125.0	125.0	125.0	125.0	125.0	125.0	125.0	125.0	125.0	125.0	125.0	139.1	139.1	148.6	150.2
92.8	92.6	92.6	93.6	95.0	95.6	88.7	86.6	99.1	99.4	103.7	103.8	112.5	108.1	113.8
87.2	87.2	87.2	86.6	86.6	86.6	86.6	86.6	86.6	86.6	86.6	88.0	86.1	87.6	98.2
71.6	74.1	75.5	76.1	77.1	73.4	70.0	75.2	77.3	81.8	82.6	89.1	90.2	100.1	100.9
68.6	68.6	68.6	68.6	69.3	69.3	74.8	74.8	74.8	74.8	75.2	75.2	86.0	86.0	86.0
56.4	55.7	55.7	55.7	57.3	57.3	57.3	57.3	57.8	57.8	57.8	71.0	71.0	78.1	78.3
69.1	69.4	69.4	69.7	69.3	69.8	76.7	76.7	76.5	76.5	76.8	77.5	78.2	82.0	82.4
91.7	86.9	89.5	89.5	89.5	89.5	89.5	95.1	95.1	95.1	95.1	103.1	104.0	104.0	104.0
75.0	75.0	69.0	69.0	70.7	70.7	59.2	62.3	62.3	63.9	63.9	71.8	71.8	71.8	71.8
82.5	82.4	82.4	82.4	82.4	82.4	83.2	83.2	84.5	91.2	91.2	99.4	101.1	101.2	108.9
48.0	48.0	49.4	49.4	49.4	49.4	49.4	49.3	49.6	49.6	54.4	54.4	55.0	55.0	55.0
72.9	72.9	72.9	72.7	71.4	72.3	73.2	72.7	75.4	75.4	75.4	82.8	82.8	91.6	91.6
48.7	48.7	48.7	48.7	48.7	48.7	58.6	58.6	57.5	58.6	58.6	58.6	58.6	62.3	62.3
85.0	85.0	89.9	89.9	89.9	89.9	89.9	92.3	92.3	92.3	89.5	89.5	89.3	89.3	95.5
66.3	66.3	65.9	65.9	65.9	65.9	65.9	67.5	70.1	77.3	77.3	77.3	77.1	77.1	77.1
178.3	178.3	178.3	165.9	166.4	166.4	167.9	167.9	167.9	168.6	168.6	168.6	168.6	168.6	191.5

Table 4-c

ESTIMATED PRICE OF NO. 2 FUEL OIL PAID BY SMALL COMMERCIAL USERS

(¢/Gal.)

Census Region SMSA	1960	1961	1962	1963	1964	1965	1966
New England							
Boston, Mass.	14.7	15.4	15.1	15.5	15.4	15.9	16.5
Middle Atlantic							
Albany, N.Y.	13.7	14.3	14.6	15.2	15.5	15.7	15.9
New York, N.Y.	14.7	15.2	15.2	15.6	15.2	15.5	15.9
Pittsburgh, Pa.	--	--	--	--	--	--	--
South Atlantic							
Atlanta, Ga.	--	--	--	--	--	--	--
Charlotte, N.C.	14.3	14.9	14.6	15.2	14.0	14.0	14.3
Washington, D.C.	14.3	15.0	15.1	15.5	15.0	15.0	15.6
Miami, Fla.	--	--	--	--	--	--	--
East North Central							
Chicago, Ill	14.3	14.6	14.9	15.1	15.0	15.2	15.3
Detroit, Mich.	15.0	15.1	15.0	15.0	15.0	15.1	15.3
Columbus, Ohio	--	--	--	--	--	--	--
East South Central							
Nashville, Tenn.	--	--	--	--	--	--	--
West North Central							
Minneapolis-St. Paul, Minn.	14.7	14.5	14.4	14.4	14.5	14.4	14.4
Omaha, Neb.	--	--	--	--	--	--	--
St. Louis, Mo.	14.8	14.9	15.1	15.2	15.3	15.3	15.5
West South Central							
Houston, Tex.	--	--	--	--	--	--	--
New Orleans, La.	--	--	--	--	--	--	--
Tulsa, Okla.	--	--	--	--	--	--	--
Mountain							
Denver, Colo.	--	--	--	--	--	--	--
Phoenix, Ariz.	--	--	--	--	--	--	--
Salt Lake City, Utah	--	--	--	--	--	--	--
Pacific							
Los Angeles, Calif.	--	--	--	--	--	--	--
San Francisco, Calif.	--	--	--	--	--	--	--
Seattle, Wash.	16.9	16.7	17.3	17.7	17.7	17.6	17.8

Source: Foster Associates, Inc.'s estimates.

1967	1968	1969	1970	1971	1972
16.9	17.2	17.5	18.6	19.6	20.1
16.7	17.8	18.1	19.1	21.1	21.4
16.5	17.0	17.4	18.3	19.6	20.0
--	--	--	--	--	--
--	--	--	--	--	--
14.3	14.7	15.6	16.3	16.4	16.4
16.3	16.8	17.3	18.3	19.1	19.2
--	--	--	--	--	--
15.6	15.9	16.2	17.0	17.7	18.0
15.9	16.3	16.8	17.5	18.1	18.3
--	--	--	--	--	--
--	--	--	--	--	--
15.2	15.7	16.2	16.9	17.5	17.6
--	--	--	--	--	--
15.9	16.4	16.9	17.6	18.6	19.0
--	--	--	--	--	--
--	--	--	--	--	--
--	--	--	--	--	--
--	--	--	--	--	--
--	--	--	--	--	--
--	--	--	--	--	--
--	--	--	--	--	--
--	--	--	--	--	--
18.3	18.8	19.5	20.3	21.6	21.7

Table 4-d

PRICES OF ELECTRICITY FOR SMALL COMMERCIAL CUSTOMERS
ELECTRICITY
($/Month)

Census Region SMSA	1960		1961		1962		1963		1964		1965	
	Jan.	July	Jan.	July	Jan.	July	Jan.	July	Jan.	July	Jan.	July
New England												
Boston, Mass.	86.07	85.78	85.78	86.59	86.59	86.31	86.31	85.93	85.93	85.93	83.88	83.88
Middle Atlantic												
Albany, N.Y.	55.77	55.70	55.70	55.70	56.52	56.55	57.10	57.11	57.11	56.88	57.29	57.29
New York, N.Y.	69.47	68.86	68.86	68.86	69.73	60.37	60.97	60.10	60.10	60.10	60.68	60.73
Pittsburgh, Pa.	53.03	53.03	53.03	53.03	53.03	53.03	53.03	53.03	53.03	53.03	53.03	53.03
South Atlantic												
Atlanta, Ga.	59.33	59.33	59.12	59.12	58.71	58.71	58.71	58.71	58.71	58.71	58.71	58.71
Charlotte, N.C.	49.44	49.44	49.44	49.44	49.44	49.44	49.44	49.44	49.44	49.44	48.41	48.41
Washington, D.C.	55.08	55.08	55.08	55.08	55.62	55.62	55.62	55.62	54.58	54.58	54.58	54.58
Miami, Fla.	89.42	89.98	64.59 1/	67.32	67.32	67.20	67.20	67.20	66.95	72.45	72.45	71.81
East North Central												
Chicago, Ill	78.14	78.14	78.14	78.14	78.14	77.84	77.84	77.63	77.63	75.31	75.31	75.10
Detroit, Mich.	69.01	69.01	69.68	69.68	69.68	69.68	69.68	69.68	69.68	69.68	72.06	72.06
Columbus, Ohio	63.24	63.24	63.24	63.24	53.77	53.77	53.77	53.77	53.77	53.77	53.77	53.77
East South Central												
Nashville, Tenn.	26.78	26.78	26.78	26.78	26.78	26.78	26.78	26.78	26.78	26.78	26.78	26.78
West South Central												
Houston, Tex.	46.90	47.40	47.40	47.76	48.72	60.16	60.16	56.17	56.17	56.17	56.17	46.17
New Orleans, La.	74.31	74.31	74.51	74.51	74.92	74.92	74.10	74.10	74.10	74.10	67.68 1/	67.68
Tulsa, Okla.	64.26	64.26	64.52	64.62	64.62	65.12	65.12	65.12	64.97	64.97	65.40	65.40
West North Central												
Minneapolis-St. Paul, Minn.	77.75	77.75	77.75	77.75	73.79	73.79	73.79	73.79	73.79	57.59	56.33	56.33
Omaha, Neb.												
Public	56.52	56.52	56.52	56.52	56.52	56.52	56.52	56.52	56.52	56.52	44.24	44.24
Private	61.00	61.30	61.30	61.00	61.00	61.30	61.00	61.00	61.00	61.00	59.06	59.06
St. Louis, Mo.	68.29	68.29	68.29	68.29	68.96	68.96	68.96	68.96	68.96	69.63	69.63	69.63
Mountain												
Denver, Colo.	60.95	60.95	65.28	65.28	65.28	65.28	65.28	65.28	65.28	65.28	65.92	65.92
Phoenix, Ariz.	55.39	55.39	55.91	55.91	55.91	55.98	55.98	55.98	55.91	55.91	55.92	55.92
Salt Lake City, Utah	52.71	52.71	52.96	52.96	52.96	52.96	53.22	57.05	57.05	57.05	57.05	57.05
Pacific												
Los Angeles, Calif.												
Public	43.96	45.63	45.63	45.63	46.45	46.45	46.45	46.45	46.25	45.84	45.84	43.86
Private	60.15	60.15	60.15	60.15	60.15	60.15	60.15	59.12	59.12	57.68	54.28	54.28
San Francisco, Calif.	61.83	61.83	61.83	61.83	61.83	61.83	61.83	61.83	61.83	60.79	60.79	60.79
Seattle, Wash.												
Public	32.45	32.45	32.45	32.45	32.45	32.45	32.45	32.45	32.45	32.45	29.97	29.97
Private	58.71	58.71	58.71	58.71	58.71	58.71	58.71	58.71	61.25	61.25	61.36	61.36

1/ New rate applies.

Note: Monthly bill represents the cost of a fixed volume based on prices in effect on January 1 and July 1 in the given year.

Source: FPC National Electric Rate Book.

1966		1967		1968		1969		1970		1971		1972		1973
Jan.	July	Jan.	July	Jan.	July	Jan.	July	Jan.	July	Jan.	July	Jan.	July	Jan.
83.75	84.88	82.00	82.00	82.00	82.00	82.00	82.00	82.00	82.00	82.00	88.77	90.10	89.27	87.44
57.29	57.29	57.83	57.83	57.83	57.83	57.83	58.30	58.30	58.30	58.85	58.85	62.35	70.13	70.57
60.73	60.46	66.43	66.43	66.43	67.79	67.79	68.06	68.06	68.06	88.31	92.95	94.95	98.64	99.22
50.55	50.55	50.55	50.55	50.55	50.55	50.87	50.87	50.87	56.20	56.20	59.38	62.55	62.95	62.76
58.71	58.71	58.71	58.71	58.71	58.71	59.99	59.99	59.99	59.99	60.83	60.83	67.63	67.83	70.95
48.41	48.41	48.41	48.41	48.41	48.41	48.41	48.41	48.41	49.48	49.48	49.48	49.48	58.21	58.21
54.57	54.57	54.58	54.58	55.11	55.11	55.10	55.10	55.10	55.11	55.11	55.11	58.02	58.13	70.22
71.81	71.29	71.29	69.85	70.53	70.16	70.16	70.21	70.20	70.32	70.32	71.08	72.97	74.55	78.78
72.49	73.09	73.09	72.11	73.69	73.69	73.69	74.20	74.20	74.77	76.49	78.80	85.04	85.63	85.36
65.52	65.66	65.66	63.49	63.49	63.49	63.49	63.68	63.68	64.41	64.41	65.35	66.90	64.84	71.40
50.99	50.99	51.23	51.23	51.48	57.92	57.92	57.92	57.92	57.92	57.92	57.92	57.92	57.92	60.26
26.78	26.78	26.78	26.78	27.89	27.89	27.89	31.11	31.11	31.11	37.18	37.18	37.18	37.18	37.18
46.17	46.32	46.32	46.32	46.77	46.77	47.45	47.45	47.45	47.45	47.85	48.58	48.63	48.63	49.58
67.07	67.07	67.07	67.07	69.05	69.05	69.05	69.05	69.71	69.07	69.07	69.07	69.71	69.49	69.49
63.53	63.53	63.35	63.35	63.65	63.65	63.70	63.61	63.61	63.61	63.72	63.72	65.50	65.50	66.23
56.33	56.33	58.02	58.02	58.02	58.02	59.70	59.70	59.70	59.70	59.70	66.31	67.97	68.23	68.75
44.70	44.70	44.90	44.86	45.98	45.98	46.25	46.31	46.42	46.70	49.16	49.16	49.64	49.78	52.91
59.06	59.06	59.06	59.06	59.64	59.64	59.64	59.64	58.50	58.50	58.50	65.88	65.61	66.41	66.68
69.63	69.63	69.63	69.63	67.66	67.66	67.66	67.66	71.90	71.90	71.90	71.90	75.10	75.10	84.22
65.92	65.92	67.84	67.84	67.84	67.84	67.84	67.84	67.84	67.84	67.84	67.84	67.84	71.38	71.38
55.98	55.98	56.15	56.15	57.39	57.39	59.23	59.32	59.32	59.32	59.56	59.56	59.94	65.62	65.62
57.05	57.05	57.05	57.05	57.05	57.05	57.60	57.34	57.34	57.34	57.34	57.34	57.34	62.31	62.31
43.86	43.96	46.52	46.41	46.41	46.41	46.41	46.41	46.41	52.47	52.47	52.47	54.12	68.49	68.69
54.28	54.28	54.28	54.28	55.34	55.34	55.34	55.34	62.10	62.10	62.10	62.10	71.42	72.08	73.65
60.79	60.79	61.97	61.97	61.97	61.97	61.97	61.97	64.33	64.33	70.31	70.31	70.69	70.69	70.69
29.97	29.97	29.97	29.97	30.05	30.05	30.05	30.05	30.20	30.20	30.20	30.20	32.42	32.42	32.42
58.41	58.41	58.41	58.41	58.38	58.38	58.38	58.38	58.38	58.38	58.38	58.38	58.38	62.67	62.67

Table 4-e

PRICES OF ELECTRICITY FOR SMALL COMMERCIAL CUSTOMERS
ELECTRICITY

(¢/Kwh)

Census Region / SMSA	1960 Jan.	1960 July	1961 Jan.	1961 July	1962 Jan.	1962 July	1963 Jan.	1963 July	1964 Jan.	1964 July	1965 Jan.	1965 July
New England												
Boston, Mass.	4.304	4.289	4.289	4.330	4.330	4.316	4.316	4.297	4.297	4.297	4.194	4.194
Middle Atlantic												
Albany, N.Y.	2.789	2.785	2.785	2.785	2.826	2.827	2.855	2.856	2.856	2.844	2.864	2.864
New York, N.Y.	3.473	3.444	3.444	3.444	3.486	3.019	3.049	3.006	3.006	3.006	3.035	3.037
Pittsburgh, Pa.	2.652	2.652	2.652	2.652	2.652	2.652	2.652	2.652	2.652	2.652	2.652	2.652
South Atlantic												
Atlanta, Ga.	2.966	2.966	2.956	2.936	2.936	2.936	2.936	2.936	2.936	2.936	2.936	2.936
Charlotte, N.C.	2.472	2.472	2.472	2.472	2.472	2.472	2.472	2.472	2.472	2.472	2.421	2.421
Washington, D.C.	2.754	2.754	2.754	2.754	2.781	2.781	2.781	2.781	2.730	2.730	2.730	2.730
Miami, Fla.	4.471	4.499	3.230	3.366	3.366	3.360	3.360	3.360	3.348	3.623	3.623	3.591
East North Central												
Chicago, Ill.	3.097	3.097	3.097	3.097	3.097	3.892	3.892	3.882	3.882	3.766	3.766	3.755
Detroit, Mich.	3.451	3.451	3.484	3.484	3.484	3.484	3.484	3.484	3.484	3.484	3.603	3.603
Columbus, Ohio	3.162	3.162	3.162	3.162	2.688	2.688	2.688	2.688	2.688	2.688	2.688	2.688
East South Central												
Nashville, Tenn.	1.339	1.339	1.339	1.339	1.339	1.339	1.339	1.339	1.339	1.339	1.339	1.339
West South Central												
Houston, Tex.	2.345	2.370	2.370	2.388	2.436	3.008	3.008	2.809	2.809	2.809	2.809	2.308
New Orleans, La.	3.716	3.716	3.726	3.726	3.746	3.746	3.706	3.706	3.706	3.706	3.384	3.384
Tulsa, Okla.	3.213	3.213	3.226	3.231	3.231	3.256	3.256	3.256	3.249	3.249	3.270	3.270
West North Central												
Minneapolis-St. Paul, Minn.	3.888	3.888	3.888	3.888	3.690	3.690	3.690	3.690	3.690	2.880	2.817	2.817
Omaha, Neb.												
Public	2.826	2.826	2.826	2.826	2.826	2.826	2.826	2.826	2.826	2.826	2.212	2.212
Private	3.050	3.065	3.065	3.050	3.050	3.065	3.050	3.050	3.035	3.035	3.035	3.035
St. Louis, Mo.	3.414	3.414	3.414	3.414	3.447	3.447	3.447	3.447	3.447	3.481	3.481	3.481
Mountain												
Denver, Colo.	3.048	3.048	3.264	3.264	3.264	3.264	3.264	3.264	3.264	3.264	3.296	3.296
Phoenix, Ariz.	2.770	2.770	2.795	2.795	2.795	2.799	2.799	2.799	2.795	2.795	2.796	2.796
Salt Lake City, Utah	2.635	2.635	2.648	2.648	2.684	2.684	2.661	2.852	2.852	2.852	2.852	2.852
Pacific												
Los Angeles, Calif.												
Public	2.198	2.281	2.281	2.281	2.323	2.323	2.323	2.323	2.312	2.292	2.292	2.193
Private	3.008	3.008	3.008	3.008	3.008	3.008	3.008	2.956	2.956	2.884	2.714	2.714
San Francisco, Calif.	3.092	3.092	3.092	3.092	3.092	3.092	3.092	3.092	3.092	3.040	3.040	3.040
Seattle, Wash.												
Public	1.622	1.622	1.622	1.622	1.622	1.622	1.622	1.622	1.622	1.622	1.498	1.498
Private	2.936	2.936	2.936	2.936	2.936	2.936	2.936	2.936	3.063	3.063	3.069	3.069

Note: Prices are those in effect on January 1 and July 1 in the given year.

Source: FPC National Electric Rate Book.

1966		1967		1968		1969		1970		1971		1972		1973
Jan.	July	Jan.	July	Jan.	July	Jan.	July	Jan.	July	Jan.	July	Jan.	July	Jan.
4.188	4.244	4.100	4.100	4.100	4.100	4.100	4.100	4.100	4.100	4.100	4.439	4.505	4.464	4.370
2.864	2.864	2.892	2.892	2.892	2.892	2.892	2.915	2.915	2.915	2.943	2.943	3.118	3.506	3.529
3.037	3.023	3.042	3.322	3.335	3.335	3.390	3.390	3.403	3.403	4.416	4.647	4.747	4.932	4.961
2.528	2.528	2.528	2.528	2.528	2.528	2.544	2.544	2.544	2.810	28.10	2.969	3.128	3.148	3.138
2.936	2.936	2.936	2.936	2.936	2.936	2.999	2.999	2.999	2.999	3.042	3.042	3.381	3.392	3.547
2.421	2.421	2.421	2.421	2.421	2.421	2.421	2.421	2.421	2.474	2.474	2.474	2.474	2.911	2.911
2.728	2.782	2.730	2.730	2.756	2.756	2.755	2.755	2.755	2.756	2.756	2.756	2.901	2.906	3.511
3.591	3.565	3.565	3.493	3.527	3.508	3.508	3.511	3.510	3.516	3.516	3.555	3.648	3.727	3.940
3.624	3.655	3.655	3.605	3.685	3.685	3.685	3.710	3.710	3.739	3.825	3.940	4.252	4.282	4.268
3.276	3.283	3.283	3.175	3.175	3.175	3.175	3.184	3.184	3.221	3.221	3.269	3.346	3.243	3.570
2.549	2.549	2.562	2.562	2.574	2.574	2.574	2.574	2.574	2.574	2.574	2.574	2.896	2.896	3.013
1.339	1.339	1.339	1.339	1.395	1.395	1.395	1.555	1.555	1.555	1.859	1.859	1.859	1.859	1.859
2.308	2.316	2.316	2.316	2.339	2.339	2.373	2.373	2.373	2.373	2.385	2.430	2.432	2.432	2.479
3.354	3.354	3.354	3.354	3.452	3.452	3.452	3.452	3.485	3.453	3.453	3.453	3.485	3.475	3.475
3.177	3.177	3.167	3.167	3.183	3.183	3.185	3.181	3.181	3.181	3.186	3.186	3.275	3.275	3.311
2.817	2.817	2.902	2.902	2.902	2.902	2.985	2.985	2.985	2.985	2.985	3.316	3.399	3.412	3.442
2.235	2.235	2.245	2.243	2.299	2.299	2.312	2.315	2.322	2.335	2.458	2.458	2.482	2.489	2.645
3.035	3.035	3.035	3.035	3.064	3.064	3.064	3.064	2.925	2.925	2.925	3.294	3.281	3.321	3.334
3.481	3.481	3.481	3.481	3.383	3.383	3.383	3.383	3.595	3.595	3.595	3.595	3.755	3.755	4.211
3.296	3.296	3.392	3.392	3.392	3.392	3.392	3.392	3.392	3.392	3.392	3.392	3.392	3.569	3.569
2.799	2.799	2.807	2.807	2.869	2.869	2.961	2.966	2.966	2.966	2.979	2.979	2.997	3.282	3.282
2.852	2.852	2.852	2.852	2.852	2.852	2.880	2.867	2.867	2.867	2.867	2.867	2.867	3.116	3.116
2.193	2.198	2.326	2.321	2.321	2.321	2.321	2.321	2.321	2.624	2.624	2.624	2.701	3.425	3.434
2.714	2.714	2.714	2.714	2.767	2.767	2.767	2.767	3.105	3.105	3.105	3.105	3.571	3.604	3.683
3.040	3.040	3.099	3.099	3.099	3.099	3.099	3.099	3.217	3.217	3.515	3.515	3.534	3.534	3.534
1.498	1.498	1.498	1.498	1.503	1.503	1.503	1.503	1.510	1.510	1.510	1.510	1.621	1.621	1.621
2.921	2.921	2.921	2.921	2.929	2.920	2.920	2.920	2.920	2.920	2.920	2.920	2.920	3.134	3.134

Table 5-a

PRICES PAID FOR FUEL BY LARGE COMMERCIAL CUSTOMERS

GAS

(¢/Mcf)

Census Region	1960		1961		1962		1963		1964		1965	
SMSA	Jan.	July	Jan.	July	Jan.	July	Jan.	July	Jan.	July	Jan.	July
New England												
Boston, Mass.	138.1	138.1	142.4	142.4	133.1	133.1	133.1	133.1	138.4	137.7	137.4	137.4
Middle Atlantic												
Albany, N.Y.	105.2	105.2	99.4	99.4	99.4	100.3	97.0	95.8	97.4	96.6	97.6	97.7
New York, N.Y.	137.3	137.3	134.3	136.0	137.4	136.1	137.7	132.8	132.9	131.3	132.3	132.0
Pittsburgh, Pa.	58.8	58.8	58.8	57.9	57.9	59.2	58.9	58.6	58.6	60.1	60.1	60.0
South Atlantic												
Atlanta, Ga.	92.7	92.7	92.7	92.7	92.7	92.7	92.7	92.7	92.7	92.7	92.7	92.7
Charlotte, N.C.	123.7	123.7	123.7	127.8	127.8	127.8	127.8	127.8	72.1 1/	72.1	72.1	72.1
Washington, D.C.	107.4	107.4	107.4	107.4	108.5	108.5	108.8	108.8	108.8	108.8	108.8	102.3
Miami, Fla.	101.6	101.6	101.6	101.6	56.9 4/	56.9	56.9	56.9	56.9	56.9	56.9	56.9
East South Central												
Nashville, Tenn.	74.1	77.4	77.4	77.4	77.4	77.4	77.4	74.1	74.1	74.1	74.1	74.1
East North Central												
Chicago, Ill.	71.0	70.9	70.9	70.8	73.1	68.7	70.6	67.0	68.4	70.9	70.3	69.8
Detroit, Mich.	74.3	74.3	77.7	77.7	77.7	77.7	73.5	73.5	73.5	73.5	73.5	73.5
Columbus, Ohio	70.0	71.0	71.5	71.0	71.0	75.0	70.0	67.0	67.5	73.0	69.0	55.2
West South Central												
Houston, Tex.	45.2	45.2	45.2	45.2	46.1	46.1	41.6	41.6	41.6	41.6	41.6	41.6
New Orleans, La.	17.4	23.6	23.6	22.7	29.0	29.0	29.0	29.0	29.0	29.0	29.0	29.0
Tulsa, Okla.	33.9	34.5	34.2	34.5	34.5	34.6	34.6	35.0	35.3	35.5	38.6	38.8
West North Central												
Minneapolis-St. Paul, Minn.	70.0	70.0	70.0	70.0	70.0	70.0	70.0	70.0	70.0	70.0	70.0	70.0
Omaha, Neb. 3/	61.3	61.3	61.3	61.3	61.3	61.3	61.3	63.5	63.5	63.5	51.0	51.0
St. Louis, Mo.	43.8	46.6	46.6	46.6	47.1	47.1	46.2	47.3	47.3	47.7	45.4	45.4
Mountain												
Denver, Colo.	30.0	33.9	44.9	44.9	44.9	44.9	44.9	44.9	44.9	44.9	45.3	45.3
Phoenix, Ariz.	48.4	48.4	49.9	49.9	50.4	50.4	46.8	46.8	46.7	49.4	46.2	47.2
Salt Lake City, Utah	31.0	32.7	32.8	32.8	32.5	32.5	32.7	32.7	32.7	32.4	32.4	32.4
Pacific												
Los Angeles, Calif.	66.2	66.2	66.2	66.2	66.2	66.2	66.2	69.0	69.0	66.8	66.8	66.8
San Francisco, Calif.	58.1	58.1	58.1	58.1	58.1	58.1	58.1	60.4	59.7	59.7	59.7	59.7
Seattle, Wash.	76.3	76.3	76.3	76.3	76.3	75.9	75.9	75.9	75.9	75.9	76.1	76.1

1/ A new rate became available for this size customer. If the old rate was used, the price decline would have been 60.6% instead of 56.4% for the large commercial and 59.1% instead of 56.5% for the industrial; hence the bulk of the change can be attributable to a price decline within existing rates rather than a change to a new rate.

2/ Unusually large purchased gas adjustment on July 1, 1971.

3/ "General Rate" up to January 1966 since no large volume commercial and industrial rate available.

4/ Change in utility.

Note: Prices are those in effect on January 1 and July 1 of the given year.

Source: American Gas Association Rate Book.

1966		1967		1968		1969		1970		1971		1972		1973
Jan.	July	Jan.	July	Jan.	July	Jan.	July	Jan.	July	Jan.	July	Jan.	July	Jan.
137.4	137.4	136.2	136.8	136.8	136.8	136.8	137.7	137.7	137.6	137.6	181.7	181.7	181.7	201.0
97.7	97.7	98.5	98.5	98.8	98.6	101.0	100.7	104.0	107.1	111.0	114.8	119.2	123.6	126.9
132.0	130.6	132.9	130.1	128.8	130.1	131.7	133.6	132.9	130.7	137.3	140.1	154.1	159.7	159.8
60.2	60.2	60.2	60.2	60.2	61.0	75.9	76.8	82.0	84.8	86.9	92.0	92.5	93.4	103.7
92.7	92.7	92.7	92.7	96.4	96.4	96.4	96.4	96.4	96.4	96.4	107.1	107.1	107.1	107.1
72.1	72.1	72.1	72.1	72.1	72.1	72.1	72.1	72.1	72.1	72.1	72.1	72.1	82.7	82.7
102.3	102.3	102.3	102.3	100.0	100.2	100.2	100.2	100.2	103.3	103.3	114.3	115.4	104.1	117.4
56.9	56.9	56.9	58.6	58.6	54.6	53.9	57.0	57.0	56.7	56.7	67.5 [2/]	56.4	56.4	56.4
74.1	74.1	74.1	74.1	74.1	74.1	74.1	74.1	74.1	74.1	74.1	88.2	88.2	94.2	98.3
69.2	68.6	68.6	69.6	70.7	71.3	64.4	62.3	74.9	75.2	79.4	79.5	88.2	83.9	89.5
73.5	73.5	73.5	73.5	73.5	73.5	73.5	73.5	73.5	73.5	77.4	77.8	75.9	77.4	88.5
60.2	62.7	63.6	63.9	64.4	63.4	59.4	62.3	63.5	67.9	72.4	73.4	75.8	88.0	88.9
41.6	41.6	41.6	41.6	42.0	42.0	47.3	47.3	47.3	47.3	47.6	47.6	63.3	63.3	63.3
29.0	27.0	27.0	27.0	27.7	27.7	27.7	27.7	28.0	28.0	28.0	36.8	36.8	37.2	39.7
38.6	38.9	38.9	39.2	38.8	39.3	41.7	41.7	41.5	41.5	41.8	42.4	42.8	46.6	47.1
70.0	69.2	71.3	71.3	71.3	71.3	71.3	76.2	76.2	76.2	76.2	91.2	92.0	92.0	92.0
57.3	42.5	42.5	42.5	44.3	44.3	44.3	45.7	45.7	46.6	46.6	53.6	53.6	53.6	53.6
45.4	45.3	45.3	45.3	45.3	45.3	45.7	45.7	48.8	55.2	55.2	63.3	64.9	64.9	72.7
45.3	45.3	46.6	46.6	46.6	46.6	46.6	46.6	47.0	47.0	48.7	48.7	49.2	49.2	49.2
48.2	48.2	48.2	48.0	46.3	47.2	47.9	47.3	50.5	50.5	50.5	57.9	57.9	63.1	63.1
32.4	32.4	32.4	32.4	32.4	32.4	35.7	35.7	35.0	35.7	35.7	35.7	35.7	39.3	39.3
64.9	64.9	68.7	68.7	68.7	68.7	68.7	70.7	70.7	70.7	74.3	74.3	74.2	74.2	80.7
59.7	59.7	58.5	58.5	58.5	58.5	58.5	59.2	61.2	65.7	65.7	65.7	65.6	65.6	65.6
76.1	76.1	76.1	76.1	76.3	76.3	77.0	77.0	77.0	77.6	77.6	77.6	77.6	77.6	90.5

Table 5-b

DELIVERED PRICES TO LARGE COMMERCIAL USERS FOR NO. 2 FUEL OIL[1/]

(Cents Per Gallon)

Census Region SMSA	1960		1961		1962		1963		1964		1965	
	Jan.	July	Jan.	July	Jan.	July	Jan.	July	Jan.	July	Jan.	July
New England												
Boston, Mass.	10.4	9.6	10.8	10.1	10.8	9.6	10.6	9.6	10.5	9.4	10.4	9.8
Middle Atlantic												
Albany, N.Y.	11.0	9.9	11.2	10.4	11.2	9.7	11.2	10.2	10.9	9.5	10.5	9.9
New York, N.Y.	11.1	10.0	11.3	10.8	11.6	9.8	11.3	10.4	10.9	9.8	10.5	10.2
Pittsburgh, Pa.	--	--	--	--	--	--	--	--	--	--	--	--
South Atlantic												
Atlanta, Ga.	--	--	--	--	--	--	--	--	--	--	--	--
Charlotte, N.C.	11.4	10.6	11.7	11.2	11.6	9.8	11.6	10.4	11.1	9.5	11.1	10.6
Washington, D.C.	11.3	10.5	11.6	10.7[2/]	11.5	9.8[3/]	11.6	10.4	11.1	9.7	10.4	10.0
Miami, Fla.	--	--	--	--	--	--	--	--	--	--	--	--
East North Central												
Chicago, Ill.	10.4	9.8	10.7	9.6	11.9	10.9	11.2	9.6	10.7	8.8	10.1	9.8
Detroit, Mich.	11.5	11.0	11.1	10.6	11.1	10.6	11.1	10.6	11.1	9.5	10.2	10.5
Columbus, Ohio	--	--	--	--	--	--	--	--	--	--	--	--
East South Central												
Nashville, Tenn.	--	--	--	--	--	--	--	--	--	--	--	--
West South Central												
Houston, Tex.	--	--	--	--	--	--	--	--	--	--	--	--
New Orleans, La.	--	--	--	--	--	--	--	--	--	--	--	--
Tulsa, Okla.	--	--	--	--	--	--	--	--	--	--	--	--
West North Central												
Minneapolis-St. Paul, Minn.	11.2	9.9	11.4	10.2	11.4	10.7	11.2	10.9	11.7	9.4	10.9	9.9
Omaha, Neb.	--	--	--	--	--	--	--	--	--	--	--	--
St. Louis, Mo.	10.4	9.7	9.7	10.2	11.3	10.8	10.8	10.3	10.3	10.0	10.7	10.7
Mountain												
Denver, Colo.	--	--	--	--	--	--	--	--	--	--	--	--
Phoenix, Ariz.	--	--	--	--	--	--	--	--	--	--	--	--
Salt Lake City, Utah	--	--	--	--	--	--	--	--	--	--	--	--
Pacific												
Los Angeles, Calif.	--	--	--	--	--	--	--	--	--	--	--	--
San Francisco, Calif.	--	--	--	--	--	--	--	--	--	--	--	--
Seattle, Wash.	11.5[e/]	11.5	11.5	11.5	11.5	12.0	12.0	11.7	11.7	11.7	11.9	11.9

1/ Includes taxes where applicable.
2/ Deduction of 0.5¢ TVA.
3/ Deduction of 1.5¢ TVA.
4/ Deduction of 0.4¢ TVA.

e/ Estimated.

Source: Prices from Platt's Oilgram Price Service, Fuel Oil and Oil Heat, oil companies; taxes from Commerce Clearing House and American Petroleum Institute; delivery charges estimated.

1966		1967		1968		1969		1970		1971		1972		1973
Jan.	July	Jan.	July	Jan.	July	Jan.	July	Jan.	July	Jan.	July	Jan.	July	Jan.
10.8	10.4	11.0	11.5	11.5	12.0	11.6	11.6	11.6	12.0	12.8	12.5	12.5	12.6	12.6
11.1	10.7	11.3	11.7	11.7	12.5	12.1	12.2	12.2	12.7	13.6	13.3	13.4	13.4	13.4
11.3	10.7	11.6	12.0	12.0	12.4	12.0	12.1	12.1	12.5	13.4	13.0	13.1	13.2	13.2
--	--	--	--	--	--	--	--	--	--	--	--	--	--	--
--	--	--	--	--	--	--	--	--	--	--	--	--	--	--
11.4	11.0	11.4	12.0	12.7	13.3	12.9	12.9	12.2	12.6	12.9	12.9	12.6	12.6	12.6
11.0	10.6	11.1	11.5	11.5	12.1	11.8[4/]	11.8[4/]	11.8[4/]	12.2	12.6[e/]	12.7	12.7	12.8	12.9
--	--	--	--	--	--	--	--	--	--	--	--	--	--	--
10.7	10.1	10.4	10.2	11.0	10.9	11.5	11.0	12.5	12.5	12.7	12.2	12.2	12.3	12.3
10.7	10.7	11.0	11.0	11.3	11.3	11.5	11.5	11.9	11.9	12.4	12.2	12.2	12.2	12.2
--	--	--	--	--	--	--	--	--	--	--	--	--	--	--
--	--	--	--	--	--	--	--	--	--	--	--	--	--	--
--	--	--	--	--	--	--	--	--	--	--	--	--	--	--
--	--	--	--	--	--	--	--	--	--	--	--	--	--	--
--	--	--	--	--	--	--	--	--	--	--	--	--	--	--
10.7	10.4	11.2	11.2	11.7	11.5	11.7	11.3	12.2	11.6	12.7	11.7	11.9	11.9	12.1
--	--	--	--	--	--	--	--	--	--	--	--	--	--	--
10.7	9.8	10.3	10.3	10.7	10.7	10.7	10.7	11.3	11.2	11.2	12.1	12.1	12.1	12.6
--	--	--	--	--	--	--	--	--	--	--	--	--	--	--
--	--	--	--	--	--	--	--	--	--	--	--	--	--	--
--	--	--	--	--	--	--	--	--	--	--	--	--	--	--
--	--	--	--	--	--	--	--	--	--	--	--	--	--	--
--	--	--	--	--	--	--	--	--	--	--	--	--	--	--
11.8	11.8	11.9	12.0	12.0	12.2	12.2	12.6	12.6	13.2[e/]	14.0	14.1	14.1	14.1	14.1

Table 5-c

DELIVERED PRICES TO LARGE COMMERCIAL USERS OF NO. 6 FUEL OIL[6/][7/]

(Dollars Per Barrel)

Census Region SMSA	1960		1961		1962		1963		1964		1965	
	Jan.	July	Jan.	July	Jan.	July	Jan.	July	Jan.	July	Jan.	July
New England												
Boston, Mass.	2.77	2.92	2.95	2.95	2.95	2.55	2.55	2.55	2.52	2.52	2.56	2.56
Middle Atlantic												
Albany, N.Y.	2.72	2.95	2.95	2.95	2.95	2.63	2.63	2.63	2.63	2.64	2.64	2.64
New York, N.Y.	2.73	2.97	2.97	2.97	2.97	2.59	2.59	2.61	2.61	2.62	2.62	2.62
Pittsburgh, Pa.	--	--	--	--	--	--	--	--	--	--	--	--
South Atlantic												
Atlanta, Ga.	--	--	--	--	--	--	--	--	--	--	--	--
Charlotte, N.C.	--	--	--	--	--	--	--	--	--	--	--	--
Washington, D.C.	2.92[e/]	3.14[e/]	3.14[e/]	3.14[e/]	3.14[e/]	2.99[e/]	2.99[e/]	2.99[e/]	2.99[e/]	3.02[e/]	3.02[e/]	3.00[e/]
Miami, Fla.	2.60	2.60	2.75	2.75	2.75	2.58	2.58	2.58	2.58	2.59	2.59	2.59
East North Central												
Chicago, Ill.	3.58[5/]	3.54[5/]	3.67[5/]	3.56[5/]	4.13[5/]	4.14[5/]	3.79[5/]	3.79[5/]	3.90[5/]	3.58[1/]	3.58[1/]	3.47[1/]
Detroit, Mich.	3.93	3.93	3.97	3.97	3.97	3.97	3.75	3.75	3.75	3.76	3.76	3.88[1/]
Columbus, Ohio	--	--	--	--	--	--	--	--	--	--	--	--
East South Central												
Nashville, Tenn.	--	--	--	--	--	--	--	--	--	--	--	--
West South Central												
Houston, Tex.	--	--	--	--	--	--	--	--	--	--	--	--
New Orleans, La.	--	--	--	--	--	--	--	--	--	--	--	--
Tulsa, Okla.	--	--	--	--	--	--	--	--	--	--	--	--
West North Central												
Minneapolis-St. Paul, Minn.	3.74	3.57	3.57	3.68	3.68	3.68	3.68	3.68	3.68	3.58	3.79	3.79
Omaha, Neb.	--	--	--	--	--	--	--	--	--	--	--	--
St. Louis, Mo.	3.33	3.33	3.33	3.33	3.20	3.20	3.20	3.36	3.36	2.94	3.10	3.10
Mountain												
Denver, Colo.	--	--	--	--	--	--	--	--	--	--	--	--
Phoenix, Ariz.	--	--	--	--	--	--	--	--	--	--	--	--
Salt Lake City, Utah.	n.a.	n.a.	n.a.	n.a.	n.a.	n.a.	n.a.	3.65	3.65	3.44	3.54	3.33
Pacific												
Los Angeles, Calif.	--	--	--	--	--	--	--	--	--	--	--	--
San Francisco, Calif.	--	--	--	--	--	--	--	--	--	--	--	--
Seattle, Wash.	3.12[2/]	3.12[2/]	3.22[2/]	3.26[2/]	3.26[2/]	3.36[2/]	3.36[2/]	3.26[2/]	3.26[2/]	3.27[2/]	3.27[2/]	3.27[2/]

1/ Sulfur - 1% max.
2/ Sulfux - 2% max.
3/ Sulfur - 0.3% max.
4/ Sulfur - 0.5% max.
5/ Low sulfur.
6/ Sulfur content is regular or no sulfur guarantee unless otherwise specified.
7/ Includes taxes where applicable.
8/ Sulfur - 0.8% max.

e/ estimated.

n.a. - not available.

Source: Prices from Platt's Oilgram Price Service and oil companies; taxes from Commercial Clearing House and American Petroleum Institute; delivery charges estimated.

1966		1967		1968		1969		1970		1971		1972		1973
Jan.	July	Jan.	July	Jan.	July	Jan.	July	Jan.	July	Jan.	July	Jan.	July	Jan.
2.56	2.56	2.53	2.53	2.56	2.57	2.57	2.15	2.28	2.89	4.39[1/]	4.60[1/]	5.16[4/][e/]	4.70[4/][e/]	5.20[4/]
2.69	2.69	2.69	2.69	2.70	2.77	2.77	2.79	2.79	3.42	4.32	4.26	3.73	3.56	3.60
2.61	2.40	2.75[1/]	2.75[2/]	2.75[2/]	2.76[2/]	2.76[2/]	2.79[2/]	2.93[1/]	3.75[1/]	4.65[1/]	4.89[1/]	5.74[3/]	5.16[3/]	5.70[3/]
--	--	--	--	--	--	--	--	--	--	--	--	--	--	--
--	--	--	--	--	--	--	--	--	--	--	--	--	--	--
--	--	--	--	--	--	--	--	--	--	--	--	--	--	--
3.00[e/]	3.00[e/]	3.00[e/]	3.01[e/]	3.01[e/]	3.08[1/]	3.11[1/]	3.14[1/]	3.32[1/]	3.43[1/]	4.94[1/]	5.16[1/]	5.16[1/]	4.87[1/]	5.23[8/]
2.59	2.59	2.59	2.59	2.59	2.60	2.60	2.61	2.61	2.95	4.35	4.36	4.06	4.08	4.30
3.57[1/]	3.41[1/]	3.61[1/]	3.57[1/]	3.67[1/]	3.79[1/]	3.80[1/]	3.69[1/]	3.92[1/]	4.37[1/]	5.03[1/]	5.15[1/]	5.16[1/]	5.17[1/]	5.17[1/]
3.88[1/]	3.88[1/]	3.88[1/]	3.88[1/]	3.88[1/]	3.89[1/]	3.89[1/]	3.89[1/]	3.89[1/]	3.90[1/]	4.48[1/]	4.48[1/]	4.92[1/]	4.92[1/]	4.92[1/]
--	--	--	--	--	--	--	--	--	--	--	--	--	--	--
--	--	--	--	--	--	--	--	--	--	--	--	--	--	--
--	--	--	--	--	--	--	--	--	--	--	--	--	--	--
--	--	--	--	--	--	--	--	--	--	--	--	--	--	--
--	--	--	--	--	--	--	--	--	--	--	--	--	--	--
3.79	3.79	3.90	3.90	3.90	3.91	3.91	4.24	4.24	3.92	4.14	4.70	4.74	4.76	4.81
--	--	--	--	--	--	--	--	--	--	--	--	--	--	--
3.10	3.10	3.10	3.10	3.10	3.11	3.11	3.33	3.33	3.37	4.36	4.37	4.37	4.21	4.39
--	--	--	--	--	--	--	--	--	--	--	--	--	--	--
--	--	--	--	--	--	--	--	--	--	--	--	--	--	--
3.33	3.17	3.37	3.49	3.49	3.50	3.64	3.31	3.64	4.10	4.64	4.65	4.65	4.67	4.67
--	--	--	--	--	--	--	--	--	--	--	--	--	--	--
--	--	--	--	--	--	--	--	--	--	--	--	--	--	--
3.27[2/]	3.27[2/]	3.27[2/]	3.28[2/]	3.28[2/]	3.29[2/]	3.29[2/]	3.30[2/]	3.32[2/]	3.57[2/]	4.57[2/]	4.89[2/]	4.79[2/]	4.81[2/]	4.82[2/]

Table 5-d
PRICES OF ELECTRICITY FOR LARGE COMMERCIAL CUSTOMERS
ELECTRICITY
(¢/Kwh)

Census Region SMSA	1960 Jan.	1960 July	1961 Jan.	1961 July	1962 Jan.	1962 July	1963 Jan.	1963 July	1964 Jan.	1964 July	1965 Jan.	1965 July
New England												
Boston, Mass.	2.072	2.057	2.057	2.098	2.098	2.084	2.084	2.065	2.065	2.065	1.966	1.966
Middle Atlantic												
Albany, N.Y.	1.192	1.189	1.189	1.189	1.197	1.199	1.211	1.211	1.211	1.200	1.212	1.204
New York, N.Y.	1.972	1.942	1.942	1.942	1.985	2.224	2.245	2.181	2.181	2.181	2.202	2.225
Pittsburgh, Pa.	1.338	1.338	1.338	1.338	1.338	1.338	1.338	1.338	1.338	1.338	1.338	1.338
South Atlantic												
Atlanta, Ga.	1.499	1.499	1.488	1.488	1.468	1.468	1.468	1.468	1.468	1.468	1.468	1.468
Charlotte, N.C.	1.112	1.112	1.112	1.112	1.112	1.112	1.112	1.112	1.112	1.112	1.098	1.098
Washington, D.C.	1.604	1.604	1.571	1.571	1.579	1.575	1.575	1.575	1.551	1.552	1.552	1.552
Miami, Fla.	1.670	1.680	1.680	1.834	1.834	1.816	1.816	1.816	1.816	1.772	1.772	1.739
East North Central												
Chicago, Ill.	1.574	1.574	1.574	1.574	1.574	1.574	1.574	1.564	1.564	1.605	1.605	1.595
Detroit, Mich.	2.049	2.042	2.062	2.062	2.055	2.055	2.034	2.034	2.031	2.031	2.396	2.396
Columbus, Ohio	1.644	1.644	1.644	1.644	1.644	1.644	1.644	1.644	1.636	1.636	1.636	1.636
East South Central												
Nashville, Tenn.	.726	.726	.726	.726	.726	.726	.726	.726	.726	.726	.726	.726
West South Central												
Houston, Tex.	1.075	1.100	1.100	1.118	1.140	1.161	1.161	1.173	1.173	1.173	1.173	1.173
New Orleans, La.	1.344	1.344	1.360	13.60	13.79	13.79	1.332	1.332	1.332	1.332	1.321	1.321
Tulsa, Okla.	1.192	1.192	1.205	1.211	1.211	1.290	1.290	1.290	1.283	1.283	1.285	1.285
West North Central												
Minneapolis-St. Paul, Minn.	1.777	1.777	1.777	1.777	1.777	1.777	1.777	1.570	1.570	1.565	1.552	1.552
Omaha, Neb.												
Public	1.056	1.090	1.090	1.059	1.059	1.049	1.049	1.049	1.056	1.056	1.014	1.014
Private	1.750	1.766	1.766	1.750	1.750	1.766	1.750	1.750	1.660	1.660	1.497	1.497
St. Louis, Mo.	1.484	1.484	1.484	1.482	1.497	1.490	1.490	1.496	1.496	1.506	1.506	1.507
Mountain												
Denver, Colo.	1.333	1.333	1.429	1.429	1.429	1.429	1.429	1.429	1.429	1.429	1.443	1.443
Phoenix, Ariz.	1.428	1.428	1.454	1.454	1.454	1.459	1.459	1.459	1.454	1.454	1.456	1.456
Salt Lake City, Utah	1.344	1.344	1.350	1.350	1.350	1.350	1.357	1.447	1.447	1.447	1.447	1.447
Pacific												
Los Angeles, Calif.												
Public	.976	.986	.986	.986	1.027	1.027	1.027	1.027	1.017	.996	.996	.936
Private	1.212	1.212	1.212	1.212	1.212	1.212	1.212	1.212	1.212	1.212	1.191	1.191
San Francisco, Calif.	1.338	1.338	1.338	1.338	1.338	1.338	1.338	1.338	1.338	1.341	1.341	1.341
Seattle, Wash.												
Public	.927	.927	.927	.927	.927	.927	.927	.927	.927	.901	.901	.901
Private	1.813	1.813	1.813	1.813	1.813	1.183	1.183	1.813	1.890	1.890	1.206 1/	1.206

1/ New rate schedule available.

Note: Prices are those in effect on January 1 and July 1 of the given year.

Source: FPC National Rate Book.

1966		1967		1968		1969		1970		1971		1972		1973
Jan.	July	Jan.	July	Jan.	July	Jan.	July	Jan.	July	Jan.	July	Jan.	July	Jan.
1.960	2.017	1.909	1.909	1.909	1.909	1.909	1.909	1.909	1.909	1.909	2.262	2.329	2.287	2.196
1.204	1.204	1.216	1.216	1.216	1.216	1.216	1.216	1.223	1.223	1.332	1.332	1.410	1.565	1.587
2.225	2.211	2.418	2.432	2.432	2.193	2.193	2.502	2.502	2.502	2.966	3.169	3.280	3.463	3.493
1.295	1.295	1.295	1.295	1.295	1.295	1.303	1.303	1.303	1.440	1.440	1.543	1.681	1.701	1.695
1.468	1.468	1.468	1.468	1.468	1.468	1.543	1.543	1.543	1.543	1.585	1.585	1.624	1.635	1.712
1.098	1.098	1.098	1.098	1.098	1.098	1.098	1.098	1.098	1.149	1.149	1.149	1.149	1.319	1.319
1.548	1.548	1.578	1.578	1.593	1.593	1.635	1.635	1.635	1.701	1.701	1.701	1.799	1.804	2.289
1.739	1.713	1.713	1.642	1,658	1.639	1.639	1.641	1.641	1.647	1.647	1.685	1.779	1.858	1.912
1.594	1.601	1.601	1.574	1.608	1.608	1.608	1.622	1.622	1.654	1.737	1.869	2.028	2.058	2.045
1.929	1.940	1.940	1.945	1.945	1.945	1.945	1.964	1.964	2.035	2.035	2.035	2.189	1.983	2.472
1.599	1.599	1.606	1.606	1.614	1.614	1.614	1.614	1.614	1.614	1.614	1.614	1.814	1.814	1.877
.726	.726	.726	.726	.770	.770	.770	.843	.843	.843	1.103	1.103	1.103	1.103	1.103
1.170	1.170	1.170	1.170	1.181	1.181	1.199	1.199	1.199	1.199	1.211	1.211	1.340	1.332	1.374
1.294	1.294	1.298	1.298	1.337	1.327	1.327	1.327	1.340	1.319	1.319	1.319	1.346	1.374	1.374
1.295	1.295	1.285	1.285	1.300	1.300	1.303	1.300	1.300	1.300	1.304	1.304	1.375	1.375	1.411
1.552	1.552	1.599	1.599	1.599	1.599	1.599	1.787	1.787	1.787	1.787	1.787	1.855	1.868	1.882
1.037	1.037	1.067	1.046	1.072	1.072	1.084	1.088	1.094	1.107	1.274	1.274	1.298	1.305	1.461
1.497	1.513	1.513	1.630	1.646	1.646	1.646	1.646	1.646	1.646	1.646	1.724	1.711	1.751	1.766
1.507	1.512	1.512	1.515	1.479	1.479	1.479	1.489	1.581	1.581	1.612	1.612	1.762	1.771	1.994
1.443	1.443	1.485	1.485	1.485	1.485	1.485	1.485	1.485	1.485	1.485	1.485	1.618	1.668	1.630
1.459	1.459	1.467	1.467	1.497	1.475	1.548	1.553	1.553	1.553	1.564	1.564	1.583	1.747	1.747
1.447	1.447	1.447	1.447	1.447	1.447	1.461	1.461	1.415	1.415	1.415	1.415	1.415	1.540	1.540
.936	.936	1.033	.991	.991	.991	.991	.991	.991	.867	.867	.867	1.126	1.462	1.471
1.191	1.191	1.191	1.191	1.214	1.214	1.214	1.214	1.269	1.269	1.269	1.269	1.474	1.507	1.585
1.341	1.341	1.367	1.367	1.367	1.367	1.367	1.367	1.419	1.419	1.505	1.505	1.533	1.533	1.533
.901	.857	.857	.859	.859	.859	.859	.859	.863	.863	.863	.863	.925	.925	.925
1.145	1.145	1.145	1.145	1.148	1.148	1.148	1.148	1.148	1.148	1.148	1.148	1.148	1.271	1.271

Table 6-a
Sheet 1 of 2
PRICES FOR FUEL PAID BY INDUSTRIAL CUSTOMERS
GAS

(¢/Mcf)

Census Region SMSA	1960 Jan.	1960 July	1961 Jan.	1961 July	1962 Jan.	1962 July	1963 Jan.	1963 July	1964 Jan.	1964 July	1965 Jan.	1965 July
New England												
Boston, Mass.												
Firm	109.7	109.7	114.0	114.0	104.7	104.7	104.7	104.7	110.0	109.3	109.0	109.0
Interruptible	(No interruptible rate available 1/)											
Middle Atlantic												
Albany, N.Y.												
Firm	101.9	101.9	95.1	95.1	95.1	95.1	91.8	90.7	92.2	91.4	91.6	91.6
Interruptible	71.9	71.9	73.0	73.0	73.0	73.0	72.2	70.5	70.1	69.3	69.5	69.6
New York, N.Y.												
Firm	127.3	127.3	129.6	131.2	132.6	131.3	131.6	127.0	127.1	125.5	125.3	125.0
Interruptible	36.0	38.4	39.4	39.4	39.4	36.2	36.2	34.0	34.0	34.0	34.0	34.0
Pittsburgh, Pa.												
Firm	51.2	51.2	51.2	50.5	50.5	51.2	50.9	50.7	50.6	51.5	51.5	51.4
Interruptible	----------											
South Atlantic												
Atlanta, Ga.												
Firm	62.9	62.9	66.6	66.6	66.6	66.6	66.6	64.4	64.4	64.4	64.4	64.4
Interruptible	24.4	24.4	27.6	27.6	27.6	27.6	27.6	27.6	27.6	28.5	28.5	28.5
Charlotte, N.C.												
Firm	121.1	121.1	121.1	125.1	125.1	125.1	125.1	125.1	70.6 3/	70.6	70.6	70.6
Interruptible	42.0	42.0	42.0	42.0	42.0	42.0	42.0	41.5	40.6	40.6	40.6	40.6
Washington, D.C.												
Firm	96.8	96.8	96.8	96.8	96.8	96.8	97.1	97.1	97.1	97.1	97.1	90.9
Interruptible	2/	2/	2/	2/	2/	56.3	56.3	56.3	56.3	56.3	56.3	54.9
Miami, Fla.												
Firm	92.7	92.7	92.7	92.7	54.3	54.3	54.3	54.3	54.3	54.3	54.3	54.3
Interruptible 8/	37.5	37.5	38.7	38.7	39.9	39.9	39.9	39.9	39.6	39.6	39.6	39.6
East South Central												
Nashville, Tenn.												
Firm	52.8	57.2	57.2	57.2	57.2	57.2	57.2	52.8	52.8	52.8	52.8	52.8
Interruptible	30.1	31.5	31.5	31.5	31.5	31.5	31.5	30.1	30.1	30.1	30.1	30.1
East North Central												
Chicago, Ill.												
Firm	64.0	63.9	63.9	63.8	66.1	61.7	63.6	60.0	61.5	63.8	63.7	63.2
Interruptible	33.3	33.3	33.3	33.3	33.3	33.3	33.3	34.1	34.1	34.1	34.1	34.1
Detroit, Mich.												
Firm	57.0	57.0	59.6	59.6	59.6	59.6	57.6	64.8	64.8	64.8	64.8	64.8
Interruptible	42.1	42.1	42.1	42.1	42.1	42.1	42.1	42.1	42.1	42.1	42.1	42.1
Columbus, Ohio												
Firm	56.8	57.8	58.3	57.8	57.8	61.8	56.8	53.8	54.3	59.8	55.8	42.0
Interruptible	(No interruptible rate available)											
West South Central												
Houston, Tex.												
Firm 6/	19.0		20.0		20.0		21.0		19.0		19.0	
Interruptible	(No interruptible rate available)											
New Orleans, La.												
Firm	14.5	19.3	19.3	19.6	24.5	24.5	24.5	24.5	24.5	24.5	24.5	24.5
Interruptible	(No interruptible rate available)											

1966		1967		1968		1969		1970		1971		1972		1973
Jan.	July	Jan.	July	Jan.	July	Jan.	July	Jan.	July	Jan.	July	Jan.	July	Jan.
109.0	109.0	107.8	134.0	134.0	134.0	134.0	134.1	134.1	134.1	134.1	179.4	179.4	179.4	198.7
---	---	---	64.3	64.3	64.3	64.3	64.5	72.1	72.1	72.1	90.7	90.7	90.7	109.6
91.6	91.5	91.5	91.5	91.8	91.6	94.4	94.2	97.3	100.2	102.9	106.4	110.5	114.7	117.8
69.5	69.6	69.5	69.5	69.7	69.5	(Interruptible rate no longer available)								
125.0	123.7	124.7	122.0	120.8	122.0	123.5	125.3	124.7	122.6	127.6	130.2	143.5	148.5	148.6
34.0	34.0	34.0	34.0	34.0	34.0	34.0	34.0	37.6	50.2	64.6	68.0	52.2[4]	52.2	52.2
51.4	51.4	51.4	51.4	51.4	51.7	53.6	54.4	57.8	59.7	61.9	67.0	67.4	68.4	77.7
(No interruptible rate)														
64.4	64.4	64.4	64.4	66.9	66.9	66.9	66.9	66.9	81.0	83.4	89.3	90.6	89.6	90.0
28.5	28.5	28.5	28.5	28.5	28.5	28.5	28.5	28.5	31.2	34.0	39.1	40.0	43.0	43.0
70.6	70.6	70.6	70.6	70.6	70.6	70.6	70.6	70.6	70.6	70.6	70.6	70.6	81.0	81.0
40.6	40.6	40.6	40.6	40.6	40.6	40.6	40.6	40.6	40.6	40.6	40.6	40.6	49.0	49.0
90.9	90.9	90.9	90.9	87.8	87.9	87.9	87.9	87.9	90.9	90.9	101.4	101.4	90.6	110.1
54.9	54.9	54.9	54.9	54.9	54.9	54.8	54.8	54.8	56.0	56.0	60.9	60.9	56.3	82.7
54.3	54.3	54.3	56.5	56.5	52.5	51.8	55.0	55.0	54.7	54.7	65.5[7]	53.7	53.7	53.7
39.6	39.6	39.6	41.2	41.2	41.2	41.2	44.1	44.1	45.1	48.3	49.8	50.4	50.4	50.4
52.8	52.8	52.8	52.8	52.8	52.8	52.8	52.8	52.8	52.8	52.8	66.4	66.4	72.1	71.7
30.1	30.1	30.1	30.1	30.1	30.1	30.1	30.1	30.1	30.1	30.1	37.2	37.2	40.6	43.0
63.2	62.5	62.5	63.5	64.6	65.1	58.3	56.1	68.7	69.0	73.2	73.4	82.0	77.7	83.4
34.1	34.5	34.5	34.5	34.8	36.1	36.1	36.1	36.1	36.5	38.0	40.3	41.8	42.5	49.3
64.8	64.8	64.8	64.8	64.8	64.8	64.8	64.8	64.8	64.8	64.8	65.2	63.5	64.8	73.8
42.1	42.1	42.1	42.1	42.1	42.1	42.1	42.1	42.1	42.1	42.1	48.4	46.7	48.0	57.6
47.1	49.5	50.5	50.7	51.2	50.3	46.3	49.2	50.2	54.6	58.3	59.3	61.6	73.8	74.7
19.0		19.0		19.0		20.0		21.0		23.0		26.0		n.a.
24.5	23.7	23.7	23.7	23.7	23.7	23.7	23.7	23.7	23.7	23.7	32.0	32.0	32.3	34.6

Table 6-a
Sheet 2 of 2
PRICES FOR FUEL PAID BY INDUSTRIAL CUSTOMERS
GAS
(¢/Mcf)

Census Region SMSA	1960 Jan.	1960 July	1961 Jan.	1961 July	1962 Jan.	1962 July	1963 Jan.	1963 July	1964 Jan.	1964 July	1965 Jan.	1965 July
West South Central (Cont.)												
Tulsa, Okla.												
Firm	23.3	23.8	23.5	23.8	23.8	24.0	24.0	24.4	24.6	24.8	25.2	25.3
Interruptible	21.3	21.8	21.5	21.8	21.8	21.9	21.9	22.2	22.4	22.8	23.3	23.4
West North Central												
Minneapolis-St. Paul, Minn.												
Firm	68.2	68.2	68.2	68.2	68.2	68.2	68.2	68.2	68.2	68.2	68.2	68.2
Interruptible	----------											
Omaha, Neb.												
Firm	------------(No industrial rate available for this volume prior to 4/6/66)--------											
Interruptible	29.0	29.0	29.0	29.0	29.0	29.0	29.0	27.5	27.5	27.5	25.8	25.8
St. Louis, Mo.												
Firm	34.0	36.6	36.6	36.6	36.6	36.8	35.8	36.8	36.8	38.3	36.0	36.0
Interruptible	5/	28.4	28.4	28.4	28.4	28.4	27.6	29.0	29.0	30.2	29.8	29.8
Mountain												
Denver, Colo.												
Firm	29.5	33.1	44.0	44.0	44.0	44.0	44.0	44.0	44.0	44.0	44.0	44.0
Interruptible	15.6	15.6	25.4	25.4	25.4	25.4	25.4	25.4	25.4	25.4	25.4	25.4
Phoenix, Ariz.												
Firm	45.0	45.0	46.5	46.5	47.0	47.0	45.8	45.8	43.4	46.0	42.9	43.8
Interruptible	----------											
Salt Lake City, Utah												
Firm	29.8	29.8	29.3	29.3	29.3	29.3	29.3	29.3	29.3	29.3	29.3	29.9
Interruptible	22.3	23.5	23.5	23.5	23.5	23.5	23.5	23.5	23.5	23.3	23.3	23.3
Pacific												
Los Angeles, Calif.												
Firm	62.6	62.6	62.6	62.6	62.6	62.6	62.6	65.2	65.2	63.1	63.1	63.1
Interruptible	40.7	40.7	40.7	40.7	40.7	40.7	40.7	40.7	42.7	42.7	41.1	41.1
San Francisco, Calif.												
Firm	55.5	55.5	55.5	55.5	55.5	55.5	55.5	57.7	57.1	57.1	57.1	57.1
Interruptible	42.5	42.5	42.5	42.5	42.5	42.5	42.5	43.2	42.5	42.5	42.5	42.5
Seattle, Wash.												
Firm	67.7	67.7	67.7	64.4	64.4	67.6	67.6	67.6	67.6	67.6	67.6	67.6
Interruptible	39.7	39.7	39.7	39.7	39.7	39.7	39.7	39.7	37.8	37.8	37.8	37.8

1/ In 1967, an optional demand rate was offered non-residential customers for prime movers and/or for boilers.
2/ No interruptible offered for volume being priced.
3/ A new rate became available for this size customer. If the old rate was used, the price decline would have been 60.6% instead of 56.4% for the large commercial and 59.1% instead of 56.5% for the industrial; hence, the bulk of the change can be attributable to a price decline within existing rates rather than a change to a new rate.
4/ New interruptible rate became available. Previous rate was tied to prices of #6 fuel oil.
5/ By special contract only.
6/ All industrial rates are negotiated. The above series is a calculated average price based on the revenue from industrial sales divided by gas deliveries to industrial customers in Mcf. The number of customers as well as the average quantity per customer varies over the period. Data are taken from annual reports and Moody's.
7/ Unusually large purchased gas adjustment on July 1, 1971.
8/ Interruptible rates are also subject to a fuel adjustment.
9/ A new interruptible rate became available. Of a decline in the retail price of 11% (i.e., from 41.1¢ to 36.6¢), 8% can be attributable to use of the new rate and 3% to a price decline.

Note: Prices are those in effect on January 1 and July 1 of the given year.

Source: American Gas Association Rate Book.

1966		1967		1968		1969		1970		1971		1972		1973
Jan.	July	Jan.	July	Jan.	July	Jan.	July	Jan.	July	Jan.	July	Jan.	July	Jan.
25.2	25.4	25.5	25.9	25.4	26.0	27.4	27.4	27.3	27.2	27.6	28.1	28.4	32.1	32.7
23.3	23.5	23.5	23.8	23.4	23.9	25.4	25.4	25.2	25.2	25.5	26.2	26.4	30.2	30.7
68.2	68.1	68.1	68.1	68.1	68.1	68.1	73.0	73.0	73.0	73.0	76.7	76.7	76.7	76.7
(No interruptible rate)														
----	38.0	38.0	38.0	39.4	39.4	39.4	40.1	40.1	41.0	41.0	48.0	48.0	48.0	48.0
27.5	27.0	27.0	26.8	26.8	26.8	27.2	27.5	27.5	28.0	28.0	32.5	32.5	35.0	35.0
36.0	35.8	35.8	35.8	35.8	35.8	36.2	36.2	38.2	44.6	44.6	52.7	54.4	54.4	62.1
29.8	29.7	29.7	29.7	29.7	29.7	30.0	30.0	31.7	36.9	36.9	45.0	43.8	42.8	47.0
44.0	44.0	44.0	44.0	44.0	44.0	44.0	44.3	44.3	44.3	45.9	45.9	46.3	46.3	46.3
25.4	25.4	25.4	25.4	25.4	25.4	25.4	25.4	25.4	25.4	27.4	27.4	27.4	27.4	27.4
44.8	44.8	44.8	44.6	42.8	43.7	43.7	43.1	46.2	46.2	46.2	53.4	53.4	58.4	58.4
(No interruptible rate available)														
29.9	29.9	29.9	29.9	29.9	29.9	29.3	29.3	29.3	29.6	29.6	29.6	29.6	33.4	33.4
23.3	23.3	23.3	23.3	23.3	23.3	23.5	23.5	23.5	23.7	23.7	23.7	23.7	27.9	27.9
61.3	61.3	61.3	61.3	61.3	61.3	61.3	61.3	61.3	61.3	66.6	66.6	66.6	66.6	72.5
41.1	41.1	41.1	36.6[9]	36.6	36.6	36.6	36.6	37.9	37.9	41.7	41.7	41.7	41.7	46.8
57.1	57.1	55.4	55.4	55.4	55.4	55.4	56.1	56.1	59.3	59.3	59.3	59.3	59.3	59.3
42.5	42.5	41.1	41.1	41.1	41.1	41.1	41.4	41.4	44.0	44.0	44.0	44.0	44.0	44.0
67.6	67.6	67.6	68.2	68.2	68.2	68.2	68.2	68.2	68.9	68.9	73.0	73.0	73.0	85.7
38.2	38.2	38.2	38.2	38.2	38.2	38.5	38.5	38.5	38.5	38.5	42.6	42.6	42.6	54.5

Table 6-b

DELIVERED PRICES TO LARGE INDUSTRIAL USERS FOR NO. 2 FUEL OIL[1/]

(Cents Per Gallon)

Census Region SMSA	1960		1961		1962		1963		1964		1965	
	Jan.	July	Jan.	July	Jan.	July	Jan.	July	Jan.	July	Jan.	July
New England												
Boston, Mass.	10.4	9.6	10.8	10.1	10.8	9.6	10.6	9.6	10.5	9.4	10.4	9.8
Middle Atlantic												
Albany, N.Y.	11.0	9.9	11.2	10.4	11.2	9.7	11.2	10.2	10.9	9.5	10.5	9.9
New York, N.Y.	10.8	9.7	11.0	10.5	11.3	9.5	11.0	10.0	10.5	9.4	10.1	9.8
Pittsburgh, Pa.	--	--	--	--	--	--	--	--	--	--	--	--
South Atlantic												
Atlanta, Ga.	--	--	--	--	--	--	--	--	--	--	--	--
Charlotte, N.C.	11.2[e/]	10.4[e/]	11.5[e/]	11.0[e/]	11.4[e/]	9.6[e/]	11.4[e/]	10.2[e/]	10.9[e/]	9.3	10.9	10.4
Washington, D.C.	11.1	10.3	11.4	10.5[2/]	11.3	9.5[3/]	11.3	10.1	10.8	9.4	10.1	9.7
Miami, Fla.	--	--	--	--	--	--	--	--	--	--	--	--
East North Central												
Chicago, Ill.	10.4	9.8	10.7	9.6	11.9	10.9	11.2	9.6	10.7	8.8	10.1	9.8
Detroit, Mich.	11.2	10.7	10.7	10.2	10.7	10.2	10.7	10.2	10.7	9.1	9.8	10.1
Columbus, Ohio	--	--	--	--	--	--	--	--	--	--	--	--
East South Central												
Nashville, Tenn.	--	--	--	--	--	--	--	--	--	--	--	--
West South Central												
Houston, Tex.	--	--	--	--	--	--	--	--	--	--	--	--
New Orleans, La.	--	--	--	--	--	--	--	--	--	--	--	--
Tulsa, Okla.	--	--	--	--	--	--	--	--	--	--	--	--
West North Central												
Minneapolis-St. Paul, Minn.	11.2	9.9	11.4	10.2	11.4	10.7	11.2	10.9	11.7	9.4	10.9	9.9
Omaha, Neb.	--	--	--	--	--	--	--	--	--	--	--	--
St. Louis, Mo.	10.2	9.5	9.5	10.0	11.0	10.5	10.5	10.0	10.0	9.7	10.4	10.4
Mountain												
Denver, Colo.	--	--	--	--	--	--	--	--	--	--	--	--
Phoenix, Ariz.	--	--	--	--	--	--	--	--	--	--	--	--
Salt Lake City, Utah	--	--	--	--	--	--	--	--	--	--	--	--
Pacific												
Los Angeles, Calif.	--	--	--	--	--	--	--	--	--	--	--	--
San Francisco, Calif.	--	--	--	--	--	--	--	--	--	--	--	--
Seattle, Wash.	11.5[e/]	11.5	11.5	11.5	11.5	12.0	12.0	11.7	11.7	11.7	11.9	11.9

1/ Includes taxes where applicable.
2/ Deduction of 0.5¢ TVA.
3/ Deduction of 1.5¢ TVA.
4/ Deduction of 0.4¢ TVA.

e/ Estimated.

Source: Prices from Platt's Oilgram Price Service, Fuel Oil and Oil Heat, oil companies; taxes from Commerce Clearing House and American Petroleum Institute; delivery charges estimated.

1966		1967		1968		1969		1970		1971		1972		1973
Jan.	July	Jan.	July	Jan.	July	Jan.	July	Jan.	July	Jan.	July	Jan.	July	Jan.
10.8	10.4	11.0	11.5	11.5	12.0	11.6	11.6	11.6	12.0	12.8	12.5	12.5	12.6	12.6
10.9	10.5	11.1	11.5	11.5	12.0	11.6	11.6	11.6	12.0	12.8	12.5	12.5	12.5	12.5
10.8	10.2	11.0	11.4	11.4	11.8	11.4	11.4	11.4	11.8	12.6	12.3	12.3	12.4	12.4
--	--	--	--	--	--	--	--	--	--	--	--	--	--	--
--	--	--	--	--	--	--	--	--	--	--	--	--	--	--
11.2	10.8	11.2	11.7	12.4	13.1	12.7	12.7	11.9	12.3	12.7	12.7	12.3	12.3	12.3
10.7	10.3	10.8	11.2	11.2	11.7	11.3[4/]	11.3[4/]	11.3[4/]	11.7[e/]	12.1[e/]	12.2	12.2	12.3	12.3
--	--	--	--	--	--	--	--	--	--	--	--	--	--	--
10.7	10.1	10.4	10.2	11.0	10.9	11.5	11.0	12.5	12.5	12.7	12.2	12.2	12.3	12.3
10.3	10.3	10.6	10.6	10.9	10.9	11.1	11.1	11.4	11.4	11.9	11.7	11.7	11.7	11.7
--	--	--	--	--	--	--	--	--	--	--	--	--	--	--
--	--	--	--	--	--	--	--	--	--	--	--	--	--	--
--	--	--	--	--	--	--	--	--	--	--	--	--	--	--
--	--	--	--	--	--	--	--	--	--	--	--	--	--	--
--	--	--	--	--	--	--	--	--	--	--	--	--	--	--
10.7	10.4	10.9	10.9	11.4	11.2	11.4	11.0	11.8	11.3	12.3	11.4	11.4	11.4	11.5
--	--	--	--	--	--	--	--	--	--	--	--	--	--	--
10.4	9.5	10.0	10.0	10.4	10.4	10.4	10.4	11.0	10.8	10.8	11.6	11.6	11.6	12.1
--	--	--	--	--	--	--	--	--	--	--	--	--	--	--
--	--	--	--	--	--	--	--	--	--	--	--	--	--	--
--	--	--	--	--	--	--	--	--	--	--	--	--	--	--
--	--	--	--	--	--	--	--	--	--	--	--	--	--	--
--	--	--	--	--	--	--	--	--	--	--	--	--	--	--
11.8	11.8	11.9	12.0	12.0	12.2	12.2	12.6	12.6	13.2[e/]	14.0	14.1	14.1	14.1	14.1

Table 6-c

DELIVERED PRICES TO LARGE INDUSTRIAL USERS OF NO. 6 FUEL OIL[1/] [2/]

(Dollars Per Barrel)

Census Region SMSA	1960 Jan.	1960 July	1961 Jan.	1961 July	1962 Jan.	1962 July	1963 Jan.	1963 July	1964 Jan.	1964 July	1965 Jan.	1965 July
New England												
Boston, Mass.	2.77	2.92	2.95	2.95	2.95	2.55	2.55	2.55	2.52	2.52	2.56	2.56
Middle Atlantic												
Albany, N.Y.	2.72	2.95	2.95	2.95	2.95	2.63	2.63	2.63	2.63	2.64	2.64	2.64
New York, N.Y.	2.65	2.88	2.88	2.88	2.88	2.51	2.51	2.51	2.51	2.52	2.52	2.52
Pittsburgh, Pa.	--	--	--	--	--	--	--	--	--	--	--	--
South Atlantic												
Atlanta, Ga.	--	--	--	--	--	--	--	--	--	--	--	--
Charlotte, N.C.	--	--	--	--	--	--	--	--	--	--	--	--
Washington, D.C.	2.86[e/]	3.08[e/]	3.08[e/]	3.08[e/]	3.08[e/]	2.90[e/]	2.90[e/]	2.90[e/]	2.90[e/]	2.93[e/]	2.93[e/]	2.91[e/]
Miami, Fla.	2.60	2.60	2.75	2.75	2.75	2.58	2.58	2.58	2.58	2.59	2.59	2.59
East North Central												
Chicago, Ill.	3.58[8/]	3.54[8/]	3.67[8/]	3.56[8/]	4.13[8/]	4.14[8/]	3.79[8/]	3.79[8/]	3.90[8/]	3.58[3/]	3.58[3/]	3.47[3/]
Detroit, Mich.	3.82	3.82	3.82	3.82	3.82	3.82	3.61	3.61	3.61	3.62	3.62	3.73[3/]
Columbus, Ohio	--	--	--	--	--	--	--	--	--	--	--	--
East South Central												
Nashville, Tenn.	--	--	--	--	--	--	--	--	--	--	--	--
West South Central												
Houston, Tex.	--	--	--	--	--	--	--	--	--	--	--	--
New Orleans, La.	--	--	--	--	--	--	--	--	--	--	--	--
Tulsa, Okla.	--	--	--	--	--	--	--	--	--	--	--	--
West North Central												
Minneapolis-St. Paul, Minn.	3.74	3.57	3.57	3.68	3.68	3.68	3.68	3.68	3.68	3.58	3.79	3.79
Omaha, Neb.	--	--	--	--	--	--	--	--	--	--	--	--
St. Louis, Mo.	3.26	3.26	3.26	3.26	3.11	3.11	3.11	3.26	3.26	2.85	3.01	3.01
Mountain												
Denver, Colo.	--	--	--	--	--	--	--	--	--	--	--	--
Phoenix, Ariz.	--	--	--	--	--	--	--	--	--	--	--	--
Salt Lake City, Utah	n.a.	n.a.	n.a.	n.a.	n.a.	n.a.	n.a.	3.53	3.53	3.32	3.42	3.22
Pacific												
Los Angeles, Calif.	--	--	--	--	--	--	--	--	--	--	--	--
San Francisco, Calif.	--	--	--	--	--	--	--	--	--	--	--	--
Seattle, Wash.	3.12[5/]	3.12[5/]	3.22[5/]	3.26[5/]	3.26[5/]	3.36[5/]	3.36[5/]	3.26[5/]	3.26[5/]	3.27[5/]	3.27[5/]	3.27[5/]

1/ Sulfur content is regular or no sulfur guarantee unless otherwise specified.
2/ Includes taxes where applicable.
3/ Sulfur - 1% max.
4/ Sulfur - 0.5% max.
5/ Sulfur - 2% max.
6/ Sulfur - 0.3% max.
7/ Sulfur - 0.8% max.
8/ Low sulfur.

n.a. - Not available.

e/ Estimated.

Source: Prices from Platt's Oilgram Price Service and oil companies; taxes from Commerce Clearing House and American Petroleum Institute; delivery charges estimated.

1966		1967		1968		1969		1970		1971		1972		1973
Jan.	July	Jan.	July	Jan.	July	Jan.	July	Jan.	July	Jan.	July	Jan.	July	Jan.
2.56	2.56	2.53	2.53	2.56	2.57	2.57	2.15	2.28	2.89	4.39[3/]	4.60[3/]	5.16[e/][4/]	4.70[e/][4/]	5.20[4/]
2.64	2.64	2.64[5/]	2.64[5/]	2.65[5/]	2.66[5/]	2.67[5/]	2.68[5/]	2.69[3/]	3.23[3/]	4.08[3/]	4.02[3/]	3.49[6/]	3.33[6/]	3.36[6/]
2.49	2.29	2.62[5/]	2.62[5/]	2.62[5/]	2.63[5/]	2.63[5/]	2.63[5/]	2.76[3/]	3.54[3/]	4.39[3/]	4.61[3/]	5.36[6/]	4.82[6/]	5.33[6/]
--	--	--	--	--	--	--	--	--	--	--	--	--	--	--
--	--	--	--	--	--	--	--	--	--	--	--	--	--	--
--	--	--	--	--	--	--	--	--	--	--	--	--	--	--
2.91[e/]	2.91[e/]	2.91[e/]	2.92[e/]	2.92[e/]	2.99[3/]	2.99[3/]	3.02[3/]	3.19[3/]	3.30[3/]	4.75[3/]	4.96[3/]	4.96[3/]	4.68[3/]	4.98[7/]
2.59	2.59	2.59	2.59	2.59	2.60	2.60	2.61	2.61	2.95	4.35	4.36	4.06	4.08	4.30
3.57[3/]	3.41[3/]	3.61[3/]	3.57[3/]	3.67[3/]	3.79[3/]	3.80[3/]	3.69[3/]	3.92[3/]	4.37[3/]	5.03[3/]	5.15[3/]	5.16[3/]	5.17[3/]	5.17[3/]
3.73[3/]	3.73[3/]	3.73[3/]	3.73[3/]	3.73[3/]	3.74[3/]	3.74[3/]	3.74[3/]	3.74[3/]	3.75[3/]	4.31[3/]	4.31[3/]	4.73[3/]	4.73[3/]	4.73[3/]
--	--	--	--	--	--	--	--	--	--	--	--	--	--	--
--	--	--	--	--	--	--	--	--	--	--	--	--	--	--
--	--	--	--	--	--	--	--	--	--	--	--	--	--	--
--	--	--	--	--	--	--	--	--	--	--	--	--	--	--
--	--	--	--	--	--	--	--	--	--	--	--	--	--	--
3.79	3.79	3.79	3.79	3.79	3.80	3.80	4.12	4.12	3.81	4.02	4.56	4.56	4.58	4.58
--	--	--	--	--	--	--	--	--	--	--	--	--	--	--
3.01	3.01	3.01	3.01	3.01	3.02	3.02	3.23	3.23	3.24	4.19	4.20	4.20	4.05	4.22
--	--	--	--	--	--	--	--	--	--	--	--	--	--	--
--	--	--	--	--	--	--	--	--	--	--	--	--	--	--
3.22	3.06	3.26	3.37	3.37	3.38	3.48	3.17	3.48	3.92	4.44	4.45	4.45	4.47	4.47
--	--	--	--	--	--	--	--	--	--	--	--	--	--	--
--	--	--	--	--	--	--	--	--	--	--	--	--	--	--
3.27[5/]	3.27[5/]	3.27[5/]	3.28[5/]	3.28[5/]	3.29[5/]	3.29[5/]	3.30[5/]	3.32[5/]	3.57[5/]	4.37[5/]	4.89[5/]	4.79[5/]	4.81[5/]	4.82[5/]

Table 6-d

PRICES FOR FUEL PAID BY INDUSTRIAL CUSTOMERS[1/][2/]

COAL

($/Ton)

Census Region SMSA	1960		1961		1962		1963		1964		1965	
	July	Dec.	July	Dec.	July	Dec.	July	Dec.	July	Dec.	July	Dec.
New England												
Boston, Mass.	--	--	--	--	--	--	--	--	--	--	--	--
Middle Atlantic												
Albany, N.Y.	--	9.74	--	9.71	--	9.77	--	9.57	--	9.60	--	9.62
New York, N.Y.	--	12.22	--	11.49	--	10.68	--	10.50	--	11.09	--	10.89
Pittsburgh, Pa.	5.82	5.82	5.82	5.82	5.82	5.82	5.82	5.82	5.82	5.82	5.92	5.92
South Atlantic												
Atlanta, Ga.[3/]	--	10.07	--	9.86	--	9.89	--	10.00	--	10.24	--	10.32
Charlotte, N.C.[3/]	--	9.18	--	8.97	--	9.00	--	9.11	--	9.35	--	9.43
Washington, D.C.[3/]	--	--	--	--	--	--	--	--	--	--	--	--
Miami, Fla.	--	--	--	--	--	--	--	--	--	--	--	--
East South Central												
Nashville, Tenn.[3/]	--	9.80	--	9.59	--	9.62	--	9.73	--	9.97	--	10.05
East North Central												
Chicago, Ill.	--	7.58	--	7.65	--	7.67	--	7.66	--	7.76	--	7.83
Detroit, Mich.	--	8.71	--	8.56	--	8.50	--	8.55	--	8.46	--	8.60
Columbus, Ohio	--	8.77	--	8.58	--	8.55	--	8.58	--	8.45	--	8.44
West South Central												
Houston, Tex.	--	--	--	--	--	--	--	--	--	--	--	--
New Orleans, La.	--	--	--	--	--	--	--	--	--	--	--	--
Tulsa, Okla.	--	--	--	--	--	--	--	--	--	--	--	--
West North Central												
Minneapolis-St. Paul, Minn.	16.45	16.45	16.10	16.10	16.10	16.35	16.35	16.65	16.65	17.05	17.05	17.05
Omaha, Neb.[3/]	--	5.84	--	6.08	--	6.23	--	6.57	--	6.79	--	7.19
St. Louis, Mo.	5.15	5.15	5.15	5.15	5.15	5.15	5.15	5.15	5.15	5.20	5.20	5.20
Mountain												
Denver, Colo.	--	6.92	--	5.70	--	5.85	--	5.76	--	5.86	--	5.85
Phoenix, Ariz.	--	--	--	--	--	--	--	--	--	--	--	--
Salt Lake City, Utah	--	5.40	--	5.64	--	5.55	--	5.71	--	5.87	7.90	7.90
Pacific												
Los Angeles, Calif.	--	--	--	--	--	--	--	--	--	--	--	--
San Francisco, Calif.	--	--	--	--	--	--	--	--	--	--	--	--
Seattle, Wash.	10.18	10.42	10.37	10.39	10.43	10.41	10.41	11.38	11.74	11.77	11.74	11.74

1/ A December price only represents an annual average.

2/ Delivered price of coal equals price at mine plus transportation plus taxes (if applicable).

3/ Partially estimated.

Source: Coal users and coal producers.

1966		1967		1968		1969		1970		1971		1972		1973
July	Dec.	July	Dec.	July	Dec.	July	Dec.	July	Dec.	July	Dec.	July	Dec.	March
--	--	--	--	--	--	--	--	--	--	--	--	--	--	--
--	9.40	--	9.84	--	10.17	--	10.48	--	11.26	--	13.97	--	15.73	16.27
--	11.50	--	11.59	--	11.71	--	12.54	--	17.36	--	20.40	--	--	--
6.12	6.12	6.12	6.12	6.12	6.42	6.75	7.10	7.48	8.83	8.83	8.98	9.83	11.03	11.03
--	10.56	--	11.29	--	11.93	--	13.13	--	17.03	--	20.02	--	21.22	22.16
--	9.68	--	10.41	--	11.06	--	12.27	--	16.21	--	19.26	--	20.42	21.39
--	--	--	--	--	--	--	--	--	--	--	--	--	--	--
--	--	--	--	--	--	--	--	--	--	--	--	--	--	--
--	10.29	--	11.02	--	11.66	--	12.86	--	16.76	--	19.75	--	20.93	21.89
--	8.04	--	8.09	--	8.50	--	8.67	--	9.48	--	11.44	--	11.86	14.42
--	8.81	--	9.14	--	9.46	--	10.16	--	12.00	--	13.57	--	14.43	15.06
--	8.30	--	8.72	--	8.23	--	7.02	--	7.65	--	8.99	--	9.01	9.59
--	--	--	--	--	--	--	--	--	--	--	--	--	--	--
--	--	--	--	--	--	--	--	--	--	--	--	--	--	--
--	--	--	--	--	--	--	--	--	--	--	--	--	--	--
17.05	17.05	18.00	18.40	18.40	18.40	18.40	20.20	23.20	24.70	27.65	27.65	27.90	27.90	28.50
--	7.49	--	7.86	--	8.17	--	8.62	--	9.01	--	9.79	--	10.21	10.65
5.24	5.24	5.28	5.29	5.29	5.60	5.75	6.07	6.13	7.44	7.44	8.68	8.68	9.52	9.60
--	6.13	--	6.18	--	6.37	--	6.72	--	7.13	--	7.65	--	8.03	8.43
--	--	--	--	--	--	--	--	--	--	--	--	--	--	--
8.17	8.17	8.17	8.17	8.32	8.32	6.49	6.62	6.62	8.47	8.87	8.87	--	9.37	9.37
--	--	--	--	--	--	--	--	--	--	--	--	--	--	--
--	--	--	--	--	--	--	--	--	--	--	--	--	--	--
11.75	11.72	12.44	12.46	12.49	13.37	13.35	14.42	15.07	16.17	16.28	17.73	17.79	19.36	19.45

Table 6-e

PRICES OF FUEL PAID BY INDUSTRIAL CUSTOMERS[1/][2/]

COAL

(¢/MMBtu)

Census Region SMSA	1960 July	1960 Dec.	1961 July	1961 Dec.	1962 July	1962 Dec.	1963 July	1963 Dec.	1964 July	1964 Dec.	1965 July	1965 Dec.
New England												
Boston, Mass.	--	--	--	--	--	--	--	--	--	--	--	--
Middle Atlantic												
Albany, N.Y.	--	36.1	--	36.2	--	34.9	--	34.4	--	34.4	--	34.4
New York, N.Y.	--	42.9	--	40.5	--	38.2	--	37.6	--	39.6	--	40.0
Pittsburgh, Pa.	21.7	21.7	21.7	22.2	22.1	21.9	21.9	22.1	21.8	22.4	22.2	22.6
South Atlantic												
Atlanta, Ga.[3/]	--	35.8	--	35.1	--	34.3	--	35.2	--	35.6	--	36.0
Charlotte, N.C.[3/]	--	32.7	--	31.9	--	31.3	--	32.1	--	32.5	--	32.9
Washington, D.C.	--	--	--	--	--	--	--	--	--	--	--	--
Miami, Fla.	--	--	--	--	--	--	--	--	--	--	--	--
East South Central												
Nashville, Tenn.[3/]	--	34.9	--	34.1	--	33.4	--	34.3	--	35.1	--	35.1
East North Central												
Chicago, Ill.	--	32.1	--	32.4	--	32.5	--	32.5	--	32.5	--	33.2
Detroit, Mich.	--	33.7	--	33.2	--	32.9	--	33.1	--	32.9	--	33.3
Columbus, Ohio	--	39.0	--	38.1	--	38.0	--	38.1	--	37.6	--	37.5
West South Central												
Houston, Tex.	--	--	--	--	--	--	--	--	--	--	--	--
New Orleans, La.	--	--	--	--	--	--	--	--	--	--	--	--
Tulsa, Okla.	--	--	--	--	--	--	--	--	--	--	--	--
West North Central												
Minneapolis-St. Paul, Minn.	57.1	57.1	55.9	55.9	55.9	56.8	56.8	57.8	57.8	59.2	59.2	59.2
Omaha, Neb.[3/]	--	22.0	--	22.9	--	23.4	--	24.7	--	25.5	--	27.0
St. Louis, Mo.	21.4	21.8	22.9	22.4	22.9	22.1	22.4	22.0	22.9	22.9	22.3	23.0
Mountain												
Denver, Colo.	--	32.5	--	26.7	--	27.4	--	27.0	--	27.5	--	27.4
Phoenix, Ariz.	--	--	--	--	--	--	--	--	--	--	--	--
Salt Lake City, Utah	--	21.1	--	21.6	--	21.9	--	22.3	--	22.9	30.9	30.9
Pacific												
Los Angeles, Calif.	--	--	--	--	--	--	--	--	--	--	--	--
San Francisco, Calif.	--	--	--	--	--	--	--	--	--	--	--	--
Seattle, Wash.	40.72	41.7	41.5	41.6	41.7	41.6	41.6	45.5	47.0	47.1	47.0	46.9

1/ A December price only represents an annual average.

2/ Delivered price of coal equals price at mine plus transportation plus taxes (if applicable).

3/ Partially estimated.

Source: Coal users and coal producers.

1966		1967		1968		1969		1970		1971		1972		1973
July	Dec.	July	Dec.	July	Dec.	July	Dec.	July	Dec.	July	Dec.	July	Dec.	Mar.
--	--	--	--	--	--	--	--	--	--	--	--	--	--	--
--	33.7	--	35.5	--	36.7	--	38.4	--	41.5	--	51.5	--	58.0	60.0
--	41.1	--	41.3	--	41.6	--	44.8	--	62.3	--	73.4	--	--	--
23.3	23.7	23.5	23.7	23.4	24.4	25.7	27.2	28.7	33.3	33.3	34.2	38.0	42.7	42.7
--	36.7	--	39.3	--	41.5	--	45.8	--	59.7	--	70.6	--	74.8	78.4
--	37.7	--	36.2	--	38.5	--	42.8	--	56.9	--	67.9	--	72.0	75.8
--	--	--	--	--	--	--	--	--	--	--	--	--	--	--
--	--	--	--	--	--	--	--	--	--	--	--	--	--	--
--	35.8	--	39.8	--	40.5	--	44.8	--	58.8	--	69.6	--	73.8	77.5
--	34.1	--	34.3	--	36.0	--	36.7	--	40.2	--	48.5	--	50.2	61.1
--	34.1	--	35.4	--	36.7	--	39.4	--	46.5	--	52.6	--	55.9	58.4
--	36.4	--	38.7	--	36.6	--	31.2	--	34.0	--	40.0	--	40.0	42.6
--	--	--	--	--	--	--	--	--	--	--	--	--	--	--
--	--	--	--	--	--	--	--	--	--	--	--	--	--	--
--	--	--	--	--	--	--	--	--	--	--	--	--	--	--
59.2	59.2	62.5	63.9	63.9	63.9	63.9	70.1	80.5	85.3	96.0	96.0	96.9	96.9	99.0
--	28.2	--	29.6	--	30.7	--	32.4	--	33.9	--	36.8	--	38.4	40.0
23.0	22.8	23.4	23.4	23.3	24.4	26.0	28.2	27.1	34.6	33.7	40.6	40.9	43.9	43.9
--	28.7	--	29.0	--	29.9	--	31.5	--	38.3	--	35.7	--	37.7	39.5
--	--	--	--	--	--	--	--	--	--	--	--	--	--	--
31.9	31.9	31.9	31.9	32.5	32.5	25.3	25.9	25.9	33.1	34.6	34.6	--	36.6	36.6
--	--	--	--	--	--	--	--	--	--	--	--	--	--	--
--	--	--	--	--	--	--	--	--	--	--	--	--	--	--
47.0	46.9	49.8	49.8	50.0	53.5	53.4	57.7	60.3	64.7	65.1	70.9	71.2	77.4	77.8

Table 6-f
PRICES OF ELECTRICITY FOR INDUSTRIAL CUSTOMERS
ELECTRICITY
(¢/Kwh)

Census Region	1960		1961		1962		1963		1964		1965	
SMSA	Jan.	July	Jan.	July	Jan.	July	Jan.	July	Jan.	July	Jan.	July
New England												
Boston, Mass.	1.218	1.204	1.204	1.244	1.244	1.230	1.230	1.212	1.212	1.212	1.205	1.205
Middle Atlantic												
Albany, N.Y.	.838	.835	.835	.835	.839	.840	.840	.841	.841	.830	.830	.824
New York, N.Y.	1.483	1.454	1.454	1.454	1.497	1.493	1.493	1.451	1.451	1.451	1.451	1.453
Pittsburgh, Pa.	.855	.855	.855	.855	.855	.855	.855	.855	.855	.855	.855	.855
South Atlantic												
Atlanta, Ga.	.822	.822	.812	.812	.792	.792	.792	.792	.792	.792	.792	.792
Charlotte, N.C.	.687	.687	.687	.687	.687	.687	.687	.687	.687	.687	.686	.686
Washington, D.C.	1.050	1.050	1.020	1.020	1.013	1.009	1.009	1.009	.959	.960	.960	.960
Miami, Fla.	1.139	1.146	1.146	1.301	1.301	1.295	1.295	1.295	1.283	1.275	1.275	1.244
East North Central												
Chicago, Ill.	.987	.987	.987	.987	.987	.987	.987	.977	.977	1.035	1.035	1.024
Detroit, Mich.	1.178	1.173	1.173	1.173	1.173	1.173	1.155	1.155	1.154	1.154	1.139	1.139
Columbus, Ohio	1.069	1.069	1.069	1.069	1.069	1.069	1.069	1.069	1.069	1.069	1.069	1.069
East South Central												
Nashville, Tenn.	.490	.490	.490	.490	.490	.490	.490	.490	.490	.490	.490	.490
West South Central												
Houston, Tex.	.660	.685	.685	.703	.703	.723	.723	.705	.705	.705	.705	.702
New Orleans, La.	.795	.795	.809	.809	.826	.826	.784	.784	.784	.784	.775	.775
Tulsa, Okla.	.856	.856	.869	.874	.874	.898	.898	.898	.895	.895	.894	.894
West North Central												
Minneapolis-St. Paul, Minn.	.980	.980	.980	.980	.980	.980	.980	.979	.979	.974	.974	.974
Omaha, Neb.												
Public	.769	.802	.802	.798	.798	.762	.762	.762	.737	.737	.737	.737
Private								1/	1.017	1.017	1.017	1.017
St. Louis	.940	.940	.940	.938	.938	.931	.931	.937	.937	.926	.926	.927
Mountain												
Denver, Colo.	.789	.789	.861	.861	.861	.861	.861	.861	.861	.861	.861	.861
Phoenix, Ariz.	.942	.942	.969	.969	.969	.971	.971	.971	.966	.966	.968	.968
Salt Lake City, Utah	.868	.868	.868	.868	.868	.868	.868	.885	.885	.885	.885	.885
Pacific												
Los Angeles, Calif.												
Public	.708	.740	.740	.740	.779	.779	.779	.779	.769	.750	.750	.698
Private	.880	.880	.880	.880	.880	.880	.880	.880	.824	.824	.815	.815
San Francisco, Calif.	.849	.849	.849	.849	.849	.849	.849	.849	.849	.848	.848	.848
Seattle, Wash.												
Public	.583	.583	.583	.583	.583	.583	.583	.583	.583	.583	.583	.583
Private	.758	.758	.758	.758	.758	.758	.758	.758	.795	.795	.795	.795

1/ No rates available for volume priced at this time or before by this utility.
2/ New rate schedules available.

Note: Prices are those in effect on January 1 and July 1 of the given year.

Source: FPC National Electric Rate Book.

1966		1967		1968		1969		1970		1971		1972		1973
Jan.	July	Jan.	July	Jan.	July	Jan.	July	Jan.	July	Jan.	July	Jan.	July	Jan.
1.198	1.255	1.176	1.176	1.176	1.176	1.176	1.176	1.176	1.176	1.176	1.507	1.572	1.531	1.442
.824	.824	.824	.824	.824	.824	.824	.824	.779	.779	.869	.869	.943	1.042	1.063
1.453	1.439	1.565	1.578	1.578	1.631	1.631	1.644	1.644	1.644	2.050	2.242	2.353	2.537	2.565
.932	.932	.932	.932	.932	.932	.938	.938	.938	1.008	1.008	1.074	1.212	1.232	1.261
.792	.792	.792	.792	.792	.792	.809	.809	.809	.809	.859	.859	.944	.949	1.003
.686	.686	.686	.686	.686	.686	.686	.686	.686	.736	.736	.736	.736	.825	.825
.998	.998	.984	.984	.984	.984	1.021	1.021	1.021	1.081	1.081	1.081	1.150	1.155	1.471
1.244	1.218	1.136	1.067	1.067	1.049	1.049	1.051	1.051	1.057	1.057	1.093	1.184	1.260	1.315
1.024	1.035	1.034	1.004	1.025	1.025	1.025	1.047	1.047	1.080	1.150	1.265	1.379	1.409	1.392
1.006	1.015	1.015	1.019	1.028	1.028	1.028	1.036	1.056	1.056	1.056	1.106	1.286	1.104	1.399
1.004	1.004	1.004	1.004	1.004	1.004	1.004	1.004	1.004	1.050	1.050	1.085	1.206	1.206	1.260
.490	.490	.490	.490	.530	.530	.530	.563	.563	.563	.765	.765	.765	.765	.765
.702	.702	.616	.616	.616	.616	.616	.616	.616	.616	.617	.617	.687	.681	.719
.755	.755	.732	.732	.732	.751	.751	.731	.731	.726	.726	.726	.756	.780	.780
.864	.864	.856	.856	.793	.793	.743	.739	.739	.739	.744	.744	.801	.801	.833
.974	.974	.974	.931	.931	.931	.931	1.058	1.058	1.058	1.058	1.114	1.164	1.176	1.201
.759	.759	.769	.767	.767	.767	.779	.790	.788	.801	.895	.895	.917	.924	1.070
1.017	1.032	1.032	1.004	1.004	.989	.989	.989	.870	.870	.920	.998	.985	1.024	1.037
.927	.933	.933	.936	.953	.953	.953	.963	1.014	1.014	1.050	1.050	1.185	1.196	1.347
.861	.861	.861	.874	.874	.874	.874	.874	.874	.874	.874	.874	.925	.948	.936
.971	.971	.979	.979	.998	.998	1.020	1.024	1.024	1.024	1.041	1.041	1.053	1.163	1.163
.885	.885	.885	.885	.885	.885	.885	.885	.885	.885	.885	.885	.885	1.002	1.002
.698	.708	.708	.698	.698	.698	.698	.698	.698	.721	.721	.721	.802	1.056	1.065
.815	.815	.815	.815	.815	.815	.815	.815	.815	.815	.815	.815	.929	.960	1.033
.848	.848	.848	.848	.848	.848	.848	.848	.848	.848	.907	.907	.936	.936	.936
.583	.583	.445[2/]	.445	.445	.445	.445	.445	.445	.445	.445	.445	.476	.476	.476
.795	.795	.795	.795	.783	.783	.783	.783	.783	.783	.783	.783	.783	.874	.874

Table 7-a
Sheet 1 of 2
PRICES OF FUEL TO POWER PLANTS OF ELECTRIC UTILITIES
COAL, OIL AND GAS

(¢/MMBtu)

Census Region SMSA		1960	1961	1962	1963	1964	1965	1966	1967	1968	1969	1970	1971	1972
New England														
Boston, Mass.	- Coal	36.5	37.3	37.0	36.1	36.8	37.0	38.7	51.2	--	--	--	--	--
	- Oil	35.0	36.6	34.8	33.8	33.0	32.5	31.9	29.4	28.4	27.7	35.8	62.4	65.2
	- Gas	--	35.0	34.0	33.7	32.1	33.6	31.2	30.1	--	--	--	--	--
Middle Atlantic														
Albany, N.Y.	- Coal	33.3	32.9	32.1	30.8	29.2	28.8	28.8	30.1	30.4	33.4	46.7	51.5	53.4
	- Oil	--	--	--	--	--	--	--	--	--	--	35.7	41.9	46.9
	- Gas	--	--	--	--	--	--	--	--	--	--	--	--	--
New York, N.Y.	- Coal	36.3	36.3	35.6	32.2	30.5	30.6	32.0	36.0	38.2	39.8	50.4	66.2	71.0
	- Oil	35.0	37.8	33.6	33.0	33.6	33.9	33.6	33.9	38.4	36.9	40.5	52.0	65.0
	- Gas	38.7	41.1	40.9	37.5	36.2	36.3	36.8	36.3	36.5	37.1	38.2	43.8	57.7
Pittsburgh, Pa.	- Coal	20.6	20.2	19.1	18.0	18.1	18.5	19.4	19.9	21.2	22.3	29.4	35.1	36.7
	- Oil	--	103.5	87.0	89.2	78.6	82.8	79.9[1]	79.3[1]	90.0[1]	96.6[1]	90.2[1]	99.7[1]	97.5[1]
	- Gas	--	--	--	--	--	--	--	--	--	--	--	--	--
South Atlantic														
Atlanta, Ga.	- Coal	29.3	28.8	29.3	28.7	28.3	28.3	28.7	28.6	29.3	31.2	38.0	42.9	44.8
	- Oil	81.2	49.3	80.2	81.1	75.0	74.3	73.9	72.9	78.3	81.3	81.6	49.8	49.9
	- Gas	27.5	24.7	--	--	--	--	--	--	26.9	26.8	29.3	33.6	35.1
Charlotte, N.C.	- Coal	26.9	26.3	26.1	26.6	26.6	26.3	26.8	27.7	27.5	29.3	40.5	44.6	43.9
	- Oil	--	--	--	--	--	--	--	--	--	--	93.1	86.6	83.8
	- Gas	--	--	--	--	--	--	--	--	--	27.1	34.8	40.2	53.8
Washington, D.C.	- Coal	31.4	34.6	27.4	28.7	29.2	28.4	29.7	31.7	33.2	35.8	49.5	56.8	54.9
	- Oil	--	--	--	--	--	--	--	--	41.3	41.1	45.2	49.5	51.3
	- Gas	--	--	--	--	--	--	--	--	--	--	--	--	--
Miami, Fla.	- Coal	--	--	--	--	--	--	--	--	--	--	--	--	--
	- Oil	34.3	39.3	32.7	33.0	32.9	32.9	32.9	32.2	31.7	29.9	31.2	44.1	55.2
	- Gas	34.9	35.6	35.5	35.0	34.0	34.5	33.9	33.3	34.1	34.8	36.4	39.8	39.3
East South Central														
Nashville, Tenn.[2]	- Coal	18.7	18.6	17.6	18.6	17.9	17.6	18.0	18.9	18.9	19.7	21.4	27.0	29.9
	- Oil	--	--	--	--	--	--	--	--	--	--	--	--	--
	- Gas	--	--	--	--	--	20.5	20.5	20.5	20.5	20.5	21.9	25.0	n.a.
East North Central														
Chicago, Ill.	- Coal	29.3	29.0	29.1	29.0	28.2	27.2	26.2	25.6	26.4	28.0	29.9	38.9	43.2
	- Oil	72.8	72.2	72.2	69.9	66.0	68.0	66.9	68.7	69.9	--	46.8	--	60.4
	- Gas	25.1	25.5	25.6	24.3	24.3	24.1	24.0	25.6	27.5	27.7	36.6	42.5	54.7

Detroit, Mich.	- Coal	30.8	30.9	29.6	29.2	29.2	28.3	28.7	29.0	29.3	29.1	33.9	39.4	43.4
	- Oil	--	--	--	--	--	--	--	--	--	59.9	56.6	67.2	68.8
	- Gas	--	--	--	--	--	--	--	--	--	46.9	48.3	54.3	59.9
Columbus, Ohio	- Coal	21.1	23.0	20.7	20.4	20.1	20.6	21.1	21.3	22.1	23.4	24.7	26.9	34.1
	- Oil	--	--	80.3	--	--	73.5	73.5	73.6	77.6	77.6	79.7	83.2	n.a.
	- Gas	--	--	--	--	--	--	--	68.0	66.0	66.0	68.9	--	--
West South Central														
Houston, Tex.	- Coal	--	--	--	--	--	--	--	--	--	--	--	--	--
	- Oil	--	--	33.2	34.0	34.0	--	32.1	26.2	--	--	--	--	43.3
	- Gas	18.7	19.7	21.6	20.0	20.0	19.7	19.7	19.7	19.7	19.8	20.3	20.6	21.3
New Orleans, La.	- Coal	--	--	--	--	--	--	--	--	--	--	--	--	--
	- Oil	30.4	--	30.4	33.5	46.2	23.7	47.1	56.2	45.7	57.2	38.6	55.5	57.2
	- Gas	18.7	23.9	23.9	21.7	21.9	19.8	19.3	19.5	19.5	19.3	20.0	20.8	21.6
Tulsa, Okla.	- Coal	--	--	--	--	--	--	--	--	--	--	--	--	--
	- Oil	28.3	27.0	58.4	29.5	28.0	55.5	--	53.3	--	57.2	58.0	62.0	60.4
	- Gas	16.9	18.7	18.4	17.8	17.9	19.0	18.5	19.8	19.5	19.3	20.2	24.4	26.7
West North Central														
Minneapolis-St. Paul, Minn.	- Coal	27.6	27.3	27.7	28.0	27.9	28.3	28.6	28.5	29.0	30.4	33.1	39.6	41.0
	- Oil	87.0	81.3	86.7	81.1	74.6	72.8	80.3	75.7	80.4	68.7	74.5	79.7	77.5
	- Gas	28.3	24.0	24.1	24.4	24.1	24.5	24.4	24.5	24.3	24.9	25.4	28.4	31.5
Omaha, Neb.														
Public	- Coal	--	20.5	31.5	30.6	29.3	29.7	30.3	31.1	30.9	31.9	34.7	39.5	42.4
	- Oil	--	--	52.6	51.9	54.1	57.6	57.9	55.1	54.2	54.6	57.8	--	79.5
	- Gas	--	--	25.9	25.9	26.1	25.7	25.3	25.3	25.7	27.0	26.7	31.0	34.2
Private	- Coal	26.6	26.4	27.4	26.8	26.1	26.0	26.9	27.6	27.6	28.7	31.7	35.2	37.8
	- Oil	66.2	67.5	70.2	66.8	70.6	69.1	66.0	69.9	67.9	74.5	77.2	82.8	84.5
	- Gas	24.7	24.6	24.8	24.8	24.7	24.8	24.6	24.5	24.5	24.7	25.2	29.2	31.2
St. Louis, Mo.	- Coal	21.2	21.0	20.1	20.5	20.6	20.6	21.1	21.3	21.7	22.5	25.0	28.3	33.1
	- Oil	54.1	51.3	48.7	48.0	46.8	46.0	46.8	46.5	47.5	42.8	50.7	59.5	52.0
	- Gas	20.3	20.4	20.5	20.4	19.2	20.4	21.3	22.8	23.0	24.2	33.1	40.6	40.3
Mountain														
Denver, Colo.	- Coal	25.1	24.4	24.2	23.6	24.1	23.8	23.7	24.0	24.0	25.8	26.2	27.8	n.a.
	- Oil	36.1	33.1	35.0	34.4	41.3	45.4	40.2	37.4	25.2[1]	25.3[1]	n.a.	n.a.	n.a.
	- Gas	22.3	22.4	26.7	21.8	21.7	21.8	21.8	21.9	22.2	22.9	24.6	26.7	n.a.
Phoenix, Ariz.	- Coal	--	--	22.7	23.7	14.4	14.3	15.4	24.0	24.6	26.5	26.8	15.8	n.a.
	- Oil	28.7	70.7	51.2	93.4	47.0	81.1	98.0	50.7	28.1	47.8	--	58.9	n.a.
	- Gas	31.8	33.5	33.4	30.3	30.1	30.3	30.2	28.1	28.4	31.7	34.4	37.0	n.a.
Salt Lake City, Utah	- Coal	21.9	21.7	22.0	22.4	22.7	22.9	21.2	21.0	19.6	20.6	22.0	23.3	25.6
	- Oil	24.1	24.8	23.6	24.9	24.0	25.0	25.0	24.8	25.0	25.2	25.1	25.5	32.2
	- Gas	26.4	26.6	26.7	26.7	26.5	26.4	26.2	26.7	26.3	26.8	26.8	28.0	31.3

(continued)

Table 7-a
Sheet 2 of 2
PRICES OF FUEL TO POWER PLANTS OF ELECTRIC UTILITIES
COAL, OIL AND GAS

(¢/MMBtu)

Census Region SMSA		1960	1961	1962	1963	1964	1965	1966	1967	1968	1969	1970	1971	1972
Pacific														
Los Angeles,														
Calif. - Public	- Coal	--	--	--	--	--	--	--	--	--	--	--	--	--
	- Oil	33.2	n.a.	36.3	34.2	29.6	30.9	32.0	30.6	30.3	38.4	35.5	43.2	73.5
	- Gas	32.0	n.a.	34.7	34.5	32.0	30.1	30.0	30.1	29.5	29.9	30.7	32.2	35.0
Private	- Coal	--	--	--	--	--	--	--	--	--	--	--	26.7	20.9
	- Oil	31.7	32.6	32.6	31.8	29.9	31.4	30.9	31.0	33.8	36.6	36.4	60.5	78.3
	- Gas	34.8	35.6	34.5	32.9	31.7	31.0	31.0	29.8	29.8	30.6	31.6	60.0	36.8
San Francisco, Calif.	- Coal	--	--	--	--	--	--	--	--	--	--	--	--	--
	- Oil	32.6	33.6	33.3	33.7	33.1	34.0	31.7	31.7	30.4	29.5	31.8	39.4	48.2
	- Gas	35.1	34.3	34.5	34.0	32.7	32.5	32.7	32.0	31.7	31.8	33.1	35.3	38.4
Seattle, Wash. Public	- Coal	--	--	--	--	--	--	--	--	--	--	--	--	--
	- Oil	31.8	31.8	31.1	30.2	n.a.	29.4	46.4	n.a.	49.5 1/	49.5 1/	76.6 1/	35.4	--
	- Gas	--	--	--	--	--	--	--	--	--	--	--	--	--
Private	- Coal	--	--	--	--	--	--	--	--	--	--	--	--	--
	- Oil	33.8	33.8	38.8	33.8	33.8	33.9	35.7	35.8	35.7	35.6	35.4	35.4	45.4
	- Gas	--	--	--	--	--	--	--	--	--	--	--	--	--

Note: Prices shown are "as burned."

1/ Estimated from f.o.b. costs.

2/ Figures for Nashville are for TVA which generates electricity for that city.

Source: National Coal Association Steam-Electric Plant Factors, Annual Reports (Forms No. 1, 1-M) filed with the Federal Power Commission.

Table 7-b

PRICES OF FUEL TO POWER PLANTS OF ELECTRIC UTILITIES
COAL
($/Ton)

Census Region SMSA	1960	1961	1962	1963	1964	1965	1966	1967	1968	1969	1970	1971	1972
New England													
Boston, Mass.	10.03	10.80	10.30	10.46	10.10	10.15	10.42	13.66	--	--	--	--	--
Middle Atlantic													
Albany, N.Y.	8.90	8.83	8.69	8.32	7.84	7.73	7.70	8.02	8.22	8.72	11.70	13.07	13.50
New York, N.Y.	9.79	9.78	9.61	8.66	8.26	8.31	8.65	9.67	10.35	10.65	12.99	16.06	16.47
Pittsburgh, Pa.	4.97	4.82	4.58	4.36	4.42	4.48	4.61	4.79	5.02	5.20	6.65	7.83	8.28
South Atlantic													
Atlanta, Ga.	7.31	7.21	7.37	7.12	7.03	7.02	6.98	6.97	7.20	7.65	9.02	9.98	10.50
Charlotte, N.C.	7.01	6.85	6.78	6.91	6.92	6.80	6.86	6.82	6.88	7.21	9.69	10.48	10.35
Washington, D.C.	8.52	9.49	8.27	7.97	7.83	7.56	7.83	8.26	8.58	9.13	12.04	13.87	13.75
Miami, Fla.	--	--	--	--	--	--	--	--	--	--	--	--	--
East South Central													
Nashville, Tenn.	4.44	4.39	4.34	4.38	4.15	4.08	4.18	4.34	4.33	4.48	4.73	5.75	6.47
East North Central													
Chicago, Ill.	6.11	6.00	6.04	6.07	5.90	5.65	5.47	5.40	5.50	5.95	6.74	7.18	8.62
Detroit, Mich.	7.54	7.45	7.15	7.08	6.93	6.84	6.84	6.91	6.94	7.03	8.05	9.30	10.44
Columbus, Ohio	4.74	5.16	4.71	4.62	4.67	4.66	4.78	4.81	5.00	5.19	5.53	5.88	7.37
West South Central													
Houston, Tex.	--	--	--	--	--	--	--	--	--	--	--	--	--
New Orleans, La.	--	--	--	--	--	--	--	--	--	--	--	--	--
Tulsa, Okla.	--	--	--	--	--	--	--	--	--	--	--	--	--
West North Central													
Minneapolis-St. Paul, Minn.	5.62	5.89	6.09	6.37	5.93	6.20	6.19	6.43	6.51	6.88	7.13	8.89	8.34
Omaha, Neb. - Public	--	5.12	7.70	7.39	7.07	7.24	3.51	7.43	7.52	7.72	8.24	8.87	9.48
- Private	5.55	5.44	5.72	5.52	5.29	5.30	5.50	5.77	5.68	5.84	6.37	6.98	7.41
St. Louis, Mo.	4.79	4.77	4.68	4.66	4.68	4.64	4.74	4.74	4.89	5.05	5.60	6.37	7.46
Mountain													
Denver, Colo.	5.11	5.14	5.72	5.04	5.11	5.04	5.04	5.06	5.08	5.43	5.60	6.14	n.a.
Phoenix, Ariz.	--	--	4.70	4.83	2.73	2.65	2.87	5.01	5.17	5.45	5.69	2.90	4.09
Salt Lake City, Utah	5.44	5.45	5.47	5.57	5.68	5.73	4.60	4.49	4.12	4.13	4.45	4.73	5.15
Pacific													
Los Angeles, Calif. - Public	--	--	--	--	--	--	--	--	--	--	--	--	--
- Private	--	--	--	--	--	--	--	--	--	--	--	5.89	4.09
San Francisco, Calif.	--	--	--	--	--	--	--	--	--	--	--	--	--
Seattle, Wash. - Public	--	--	--	--	--	--	--	--	--	--	--	--	--
- Private	--	--	--	--	--	--	--	--	--	--	--	--	--

Note: Figures for Nashville are for TVA which generates the electricity for that city. Prices are "as burned."

Source: National Coal Association Steam-Electric Plant Factors, Annual Reports (Forms No. 1, 1-M) filed with the Federal Power Commission.

Table 7-c

PRICES OF FUEL TO POWER PLANTS OF ELECTRIC UTILITIES
OIL
($/Bbl.)

Census Region SMSA	1960	1961	1962	1963	1964	1965	1966	1967	1968	1969	1970	1971	1972
New England													
Boston, Mass.	2.20	2.30	2.18	2.12	2.07	2.04	1.99	1.84	1.78	1.73	2.23	3.86	3.99
Middle Atlantic													
Albany, N.Y.	--	--	--	--	--	--	--	--	--	--	2.10	2.65	2.95
New York, N.Y.	2.21	2.37	2.13	2.06	2.10	2.12	2.05[1]	2.10[1]	2.37[1/2]	2.27[1/2]	2.49[1/2]	3.18[1]	9.39[1]
Pittsburgh, Pa.	--	5.81	4.89	5.02	4.37	4.66	4.49[1]	4.65[1]	5.07[1/2]	5.43[1/2]	5.08[1/2]	5.73[1]	5.77[1]
South Atlantic													
Atlanta, Ga.	4.63	2.92	4.58	4.63	4.28	4.24	4.23	4.16	4.47[2]	4.64[2]	4.66[2]	3.03	2.87
Charlotte, N.C.	--	--	--	--	--	--	--	--	--	--	5.46[2]	5.04	4.87
Washington, D.C.	n.a.	n.a.	n.a.	n.a.	n.a.	n.a.	n.a.	n.a.	2.56	2.54	2.69	3.03	3.17
Miami, Fla.	2.19	2.13	2.17	2.09	2.08	2.07	2.07	2.03	1.99	1.88	1.96	2.74	3.40
East South Central													
Nashville, Tenn.	--	--	--	--	--	--	--	--	--	--	--	--	--
East North Central													
Chicago, Ill.	4.20	4.24	4.25	4.05	3.83	3.95	3.88	3.99	4.06[2]	--	2.94	--	3.78
Detroit, Mich.	--	--	--	--	--	--	--	--	--	3.51[2]	3.50[2]	4.09	4.22
Columbus, Ohio	--	--	4.62	--	--	4.23	4.23	4.24	4.40[2]	4.46[2]	4.58[2]	4.79	4.93
West South Central													
Houston, Tex.	--	--	2.12	2.16	2.16	--	2.05	1.63	--	--[2]	--	--	2.71
New Orleans, La.	1.90	--	1.90	2.08	2.91	1.42	2.96	3.41	2.85	3.34[2]	2.41[2]	3.46	3.63
Tulsa, Okla.	1.67	1.71	3.73	1.89	1.84	3.24	--	3.36	--	3.34[2]	3.37[2]	3.60	3.63
West North Central													
Minneapolis-St. Paul, Minn.	4.90	4.58	4.89	4.58	4.20	4.10	4.53	4.28	4.54[2]	3.88	4.32	4.82	4.74
Omaha, Neb. - Public	--	--	3.36	3.28	3.44	3.65	3.69	3.43	3.45	3.47	3.64	--	4.79
- Private	3.90	3.97	4.13	4.37	4.18	4.06	3.88	3.96	4.00	4.39	4.47	4.75	4.89
St. Louis, Mo.	3.45	3.28	3.18	3.06	3.00	2.94	2.98	2.93	2.97	2.71	3.19	3.72	3.25
Mountain													
Denver, Colo.	2.26	2.10	2.22	2.18	2.60	2.78	2.50	2.34	1.61[1]	1.62[1]	n.a.	n.a.	n.a.
Phoenix, Ariz.	1.83	4.70	3.01	4.19	2.89	4.68	5.65	2.95	1.87	3.09	--	3.75	5.54
Salt Lake City, Utah	1.54	2.15	1.51	1.60	1.58	1.61	1.62	1.62	1.63	1.63	1.63	1.64	2.43
Pacific													
Los Angeles, Calif. - Public	2.12	n.a.	2.32	2.20	1.90	1.98	2.04	1.96	1.92	2.41	2.23	2.70	4.56
- Private	2.02	2.10	2.10	2.05	1.92	2.00	1.97	1.97	2.13	2.24	2.35	3.69	4.80
San Francisco, Calif.	2.07	2.14	2.13	2.14	2.10	2.16	2.00	2.02	1.93[1]	1.86[1]	2.00[1]	2.47	3.03
Seattle, Wash. - Public	2.02	2.02	1.97	1.92	n.a.	1.89	2.98	2.92	3.19[1]	3.19[1]	4.93[1]	2.34	--
- Private	2.24	2.24	2.24	2.24	2.24	2.24	2.36	2.37	2.36	2.36	2.34	2.34	2.93

Note: Prices are "as burned" for residual fuel oil but include some distillate fuel oil.

1/ "As burned" cost not available; f.o.b. plant cost used.
2/ Distillate oil only.

Source: National Coal Association Steam-Electric Plant Factors, Annual Reports (Forms No. 1, 1-M) filed with the Federal Power Commission.

Table 7-d

PRICES OF FUEL TO POWER PLANTS OF ELECTRIC UTILITIES
GAS
(¢/Mcf)

Census Region SMSA	1960	1961	1962	1963	1964	1965	1966	1967	1968	1969	1970	1971	1972
New England													
Boston, Mass.	--	35.0	34.0	33.7	32.1	33.6	31.2	30.1	--	--	--	--	--
Middle Atlantic													
Albany, N.Y.	--	--	--	--	--	--	--	--	--	--	--	--	--
New York, N.Y.	40.9	49.3	42.7	36.4	37.9	37.8	38.3	37.7	37.8	38.4	39.5	45.2	59.4
Pittsburgh, Pa.	--	--	--	--	--	--	--	--	--	--	--	--	--
South Atlantic													
Atlanta, Ga.	28.4	25.6	--	--	--	--	--	--	28.0	28.0	30.4	34.8	36.9
Charlotte, N.C.	26.7	--	--	--	--	--	--	--	--	28.0	36.0	41.3	55.4
Washington, D.C.	--	--	--	--	--	--	--	--	--	--	--	--	--
Miami, Fla.	34.9	35.6	35.5	35.0	34.0	34.5	33.9	33.3	34.1	34.8	36.4	39.8	39.3
East South Central													
Nashville, Tenn.	--	--	--	--	--	21.5	21.5	21.5	21.5	21.5	23.0	26.3	29.0
East North Central													
Chicago, Ill.	26.5	26.7	26.7	25.3	25.3	25.3	25.3	26.7	28.7	28.5	38.2	44.0	56.3
Detroit, Mich.	--	--	--	--	--	--	--	--	--	47.9	49.4	55.2	61.1
Columbus, Ohio	--	--	--	--	--	--	--	70.0	68.0	68.0	71.0	--	--
West South Central													
Houston, Tex.	19.6	20.7	22.6	20.9	21.0	20.8	20.7	20.7	20.5	20.5	21.0	21.3	22.1
New Orleans, La.	20.3	25.7	25.8	23.4	23.6	21.3	21.0	21.2	21.2	20.8	21.4	21.8	22.9
Tulsa, Okla.	17.8	19.5	19.3	18.9	19.2	20.4	19.5	20.4	20.2	20.0	20.8	26.0	28.8
West North Central													
Minneapolis-St. Paul, Minn.	28.4	24.4	24.4	24.2	24.1	24.4	24.5	24.8	24.5	25.1	25.4	28.3	31.7
Omaha, Neb. - Public	n.a.	n.a.	26.4	25.8	26.0	25.7	25.6	25.7	26.2	27.2	26.8	31.0	34.2
- Private	24.9	25.0	25.0	24.6	24.7	24.7	25.2	24.7	24.8	24.9	25.3	29.2	31.4
St. Louis, Mo.	21.5	21.5	21.5	21.5	20.1	21.3	22.2	23.8	23.9	25.1	34.2	42.0	41.7
Mountain													
Denver, Colo.	18.8	18.8	18.5	18.2	18.2	18.2	18.4	18.6	18.7	19.3	20.7	22.6	n.a.
Phoenix, Ariz.	34.3	35.8	35.7	33.9	32.8	32.7	32.6	30.0	30.7	33.7	37.2	39.6	41.0
Salt Lake City, Utah	24.0	24.4	24.9	24.8	24.7	24.7	24.3	25.1	24.6	25.0	25.1	26.0	29.3
Pacific													
Los Angeles, Calif. - Public	35.0	n.a.	37.8	37.1	34.3	32.4	32.4	32.3	31.6	31.9	32.8	34.2	36.8
- Private	36.3	38.2	37.4	35.5	34.1	33.4	33.4	32.0	32.1	32.6	33.6	36.1	38.9
San Francisco, Calif.	36.8	37.6	37.6	36.2	34.9	34.9	35.0	34.3	34.1	34.2	35.4	37.7	41.0
Seattle, Wash. - Public	--	--	--	--	--	--	--	--	--	--	--	--	--
- Private	--	--	--	--	--	--	--	--	--	--	--	--	--

Note: Figures for Nashville are for TVA which generates the electricity for that city.
Prices are "as burned."

Source: National Coal Association Steam-Electric Plant Factors, Annual Reports (Forms No. 1, 1-M), filed with the Federal Power Commission.

Table 8-a
WHOLESALE PRICE OF GAS
(¢/MMBtu)

Census Region SMSA	1960	1961	1962	1963	1964	1965	1966	1967	1968	1969	1970	1971	1972
New England													
Boston, Mass.	57.4	56.9	57.2	57.4	56.2	56.7	60.1	58.5	60.0	63.9	66.9	71.4	75.4
Middle Atlantic													
Albany, N.Y. 3/	55.0	55.0	53.7	51.4	52.0	51.8	50.7	52.1	52.1	52.9	56.1	62.5	68.9
New York, N.Y.	42.1	43.7	45.0	44.4	43.9	44.4	43.9	43.2	42.3	42.4	43.9	48.2	53.7
Pittsburgh, Pa.	34.9	35.6	35.5	35.5	35.7	35.7	33.2	33.3	33.5	34.3	34.4	40.2	42.4
South Atlantic													
Atlanta, Ga.	35.7	39.1	37.9	38.0	36.4	36.0	35.0	34.5	33.8	33.0	37.7	46.7	46.8
Charlotte, N.C.	33.0	36.8	36.2	36.5	36.0	35.3	34.2	34.2	34.3	34.0	35.0	39.4	44.0
Washington, D.C.	54.8	57.0	57.0	57.5	51.9	48.7	47.2	47.5	45.9	46.5	48.0	56.2	58.9
Miami, Fla.	57.7	54.4	50.1	48.1	47.6	48.1	48.4	47.5	46.5	48.7	50.4	55.8	55.2
East North Central													
Chicago, Ill.	31.7	32.2	31.7	31.1	30.4	29.6	29.2	29.1	29.6	29.7	31.8	36.2	40.9
Detroit, Mich.	36.9	38.3	38.5	37.8	37.0	35.6	36.0	36.5	36.9	38.5	39.8	41.8	47.5
Columbus, Ohio 4/	46.6	46.5	46.8	46.7	45.8	44.3	43.8	43.9	41.5	42.6	44.3	45.6	48.0
East South Central													
Nashville, Tenn.	40.1	39.1	43.5	38.3	35.8	37.4	35.4	34.4	33.9	33.0	36.0	43.4	44.5
West South Central													
Houston, Tex. 2/	22.3	25.0	25.3	25.1	25.0	23.1	23.9	25.2	25.8	26.9	29.3	33.0	35.4
New Orleans, La. 5/	17.6	19.3	23.7	23.8	23.7	23.6	23.7	23.7	23.7	23.7	23.7	27.9	34.5
Tulsa, Okla. 1/	12.5	13.8	14.1	14.6	15.1	15.9	16.1	16.4	16.7	17.1	17.4	17.7	20.4
West North Central													
Minneapolis-St. Paul, Minn.	45.0	42.9	42.8	40.2	38.5	36.5	37.3	34.8	35.7	36.5	36.6	42.1	45.3
Omaha, Neb.	35.0	33.3	34.4	34.4	33.6	32.7	31.9	31.2	31.7	33.5	33.8	38.6	41.8
St. Louis, Mo.	29.6	36.2	33.1	33.3	34.0	32.9	32.2	32.4	32.6	32.6	37.3	43.7	47.0
Mountain													
Denver, Colo.	27.5	27.4	28.2	28.4	26.4	26.2	26.2	26.3	26.2	25.8	26.9	28.9	32.9
Phoenix, Ariz.	32.8	34.4	34.2	32.3	31.3	31.1	31.0	30.0	30.1	32.4	35.3	38.7	41.1
Salt Lake City 6/	23.5	23.3	25.6	24.9	25.1	25.0	24.6	23.9	23.1	23.1	24.1	30.4	36.5
Pacific													
Los Angeles, Calif.	29.3	32.0	30.7	29.0	27.1	27.1	26.6	26.6	26.3	29.2	32.1	35.6	37.6
San Francisco, Calif.	29.4	30.9	32.7	32.1	29.3	28.8	28.1	26.9	27.0	28.1	30.7	32.1	36.6
Seattle, Wash.	31.9	33.5	41.1	35.0	34.8	34.4	33.6	32.9	32.1	32.5	33.2	35.7	41.3

Note: Data are obtained from Form 2 Annual Reports to the Federal Power Commission unless noted otherwise.

1/ Data from Annual Report to Stockholders since suppliers are intrastate and do not report to FPC.
2/ Data obtained from company.
3/ Data before 1965 obtained from shareholders Annual Reports and Moody's Public Utility Manual.
4/ Prices imputed 1960-1963 since Ohio Fuel Gas was both supplier and distributor.
5/ Data 1960-1970 from company.
6/ Price paid to interstate pipelines only. Purchase price from local producers includes field sales.

Table 8-b

IMPUTED WHOLESALE PRICE OF ELECTRICITY

(¢/Kwh)

Census Region SMSA	1960	1961	1962	1963	1964	1965	1966	1967	1968	1969	1970	1971	1972
New England													
Boston, Mass.	1.05	1.12	1.07	1.04	1.10	1.03	.92	.89	.83	.86	.96	1.26	1.58
Middle Atlantic													
Albany, N.Y.	.83	.91	.88	.78	.75	.71	.71	.72	.73	.91	1.01	1.00	1.02
New York, N.Y.	1.29	1.33	1.38	1.25	1.22	1.30	1.24	1.25	1.27	1.29	1.47	1.62	1.84
Pittsburgh, Pa.	.86	.82	.79	.76	.74	.76	.74	.80	.76	.77	.99	1.05	1.07
South Atlantic													
Atlanta, Ga.	.68	.70	.72	.72	.69	.67	.63	.65	.64	.65	.72	.83	.88
Charlotte, N.C.	.58	.59	.56	.59	.55	.55	.55	.55	.54	.56	.69	.75	.79
Washington, D.C.	.82	.79	.81	.78	.83	.75	.75	.75	.80	.87	1.08	1.16	1.05
Miami, Fla.	.80	.79	.76	.73	.70	.71	.67	.67	.64	.64	.65	.77	.92
East North Central													
Chicago, Ill.	.93	.90	.91	.91	.84	.82	.80	.80	.81	.85	.93	1.04	1.10
Detroit, Mich.	.88	.92	.82	.77	.75	.74	.78	.81	.82	.83	.94	1.06	1.10
Columbus, Ohio	.88	.86	.91	.84	.84	.81	.84	.83	.84	.88	.89	.93	.96
East South Central 1/													
Nashville, Tenn. 2/	.41	.41	.41	.42	.40	.41	.40	.41	.45	.46	.53	.68	.72
West South Central													
Houston, Tex.	.54	.57	.54	.49	.48	.44	.46	.46	.46	.45	.49	.47	.49
New Orleans, La.	.73	.87	.86	.84	.80	.70	.66	.67	.56	.51	.51	.54	.63
Tulsa, Okla.	.65	.75	.66	.60	.58	.58	.54	.60	.54	.50	.50	.52	.53
West North Central													
Minneapolis-St. Paul, Minn.	.86	.81	.78	.77	.77	.82	.82	.86	.85	.87	.89	1.04	.98
Omaha, Neb. - Public	.77	.76	.73	.71	.72	.69	.68	.66	.61	.67	.66	.73	.76
- Private	.99	.99	.98	.95	1.00	.98	.94	.93	.95	.95	.96	.95	1.03
St. Louis, Mo.	.79	.85	.81	.85	.85	.81	.89	.94	.88	.83	.93	1.04	1.03
Mountain													
Denver, Colo.	.87	.82	.90	.83	.85	.83	.78	.84	.90	.88	.82	.86	.82
Phoenix, Ariz.	.81	.76	.76	.77	.69	.64	.69	.64	.65	.66	.63	.64	.84
Salt Lake City, Utah	.83	.83	.90	.95	.92	.83	.77	.80	.90	.78	.77	.88	.82
Pacific													
Los Angeles, Calif. - Public 1/	.66	.68	.67	.68	.62	.62	.59	.61	.58	.59	.60	.64	.77
- Private	.74	.75	.72	.71	.72	.69	.70	.69	.70	.71	.72	.87	.92
San Francisco, Calif.	.81	.82	.80	.83	.81	.83	.79	.81	.79	.80	.81	.83	.88
Seattle, Wash. -													
Public - generated	.59	.53	.62	.54	.50	.60	.57	.56	.36	.31	.39	.40	.39
- purchased 3/	.20	.22	.20	.21	.20	.20	.21	.21	.22	.23	.29	.27	.27
Private	.94	.85	1.06	1.05	1.11	1.13	1.10	1.09	1.04	1.06	1.06	.94	1.13

1/ Fiscal year ending June.
2/ Actual purchases from TVA.
3/ Actual purchases from Bonneville Power Administration.

Source: Imputations based on data taken from Forms 1 and 1-M Annual Reports of Electric Utilities to the FPC.

Table 8-c

WHOLESALE PRICE FOR NO. 2 FUEL OIL - FOB TERMINAL

(Cents Per Gallon)

Census Region SMSA	1960 Jan.	1960 July	1961 Jan.	1961 July	1962 Jan.	1962 July	1963 Jan.	1963 July	1964 Jan.	1964 July	1965 Jan.	1965 July
New England												
Boston, Mass.	9.9	9.1	10.3	9.6	10.3	9.1	10.1	9.1	10.0	8.9	9.9	9.3
Middle Atlantic												
Albany, N.Y.	10.5	9.4	10.7	9.9	10.7	9.2	10.7	9.7	10.4	9.0	10.0	9.4
New York, N.Y.	10.3	9.2	10.5	10.0	10.8	9.0	10.5	9.5	10.0	8.8	9.5	9.2
Pittsburgh, Pa.	--	--	--	--	--	--	--	--	--	--	--	--
South Atlantic												
Atlanta, Ga.	--	--	--	--	--	--	--	--	--	--	--	--
Charlotte, N.C.	10.6 e/	9.8 e/	10.9 e/	10.4 e/	10.8 e/	9.0 e/	10.8 e/	9.6 e/	10.3 e/	8.7	10.3	9.8
Washington, D.C.	10.6	9.8	10.9	10.0 1/	10.8	9.0 2/	10.8	9.6	10.3	8.9	9.6	9.2
Miami, Fla.	--	--	--	--	--	--	--	--	--	--	--	--
East North Central												
Chicago, Ill.	9.6	9.0	9.8	8.8	11.0	10.0	10.3	8.8	9.8	8.0	9.3	9.0
Detroit, Mich.	10.8	10.3	10.3	9.8	10.3	9.8	10.3	9.8	10.3	8.6	9.4	9.6
Columbus, Ohio	--	--	--	--	--	--	--	--	--	--	--	--
East South Central												
Nashville, Tenn.	--	--	--	--	--	--	--	--	--	--	--	--
West South Central												
Houston, Tex.	--	--	--	--	--	--	--	--	--	--	--	--
New Orleans, La.	--	--	--	--	--	--	--	--	--	--	--	--
Tulsa, Okla.	--	--	--	--	--	--	--	--	--	--	--	--
West North Central												
Minneapolis-St. Paul, Minn.	10.7	9.4	10.9	9.7	10.9	10.2	10.7	10.4	11.2	8.9	10.4	9.4
Omaha, Neb.	--	--	--	--	--	--	--	--	--	--	--	--
St. Louis, Mo.	9.8	9.0	9.0	9.5	10.5	10.0	10.0	9.5	9.5	9.3	9.9	9.9
Mountain												
Denver, Colo.	--	--	--	--	--	--	--	--	--	--	--	--
Phoenix, Ariz.	--	--	--	--	--	--	--	--	--	--	--	--
Salt Lake City, Utah.	--	--	--	--	--	--	--	--	--	--	--	--
Pacific												
Los Angeles, Calif.	--	--	--	--	--	--	--	--	--	--	--	--
San Francisco, Calif.	--	--	--	--	--	--	--	--	--	--	--	--
Seattle, Wash.	10.6 e/	10.6	10.6	10.6	10.6	11.0	11.0	10.7	10.7	10.7	10.9	10.9

1/ Deduction of 0.5¢ TVA.
2/ Deduction of 1.5¢ TVA.
3/ Deduction of 0.4¢ TVA.

e/ Estimated.

Source: Platt's Oilgram Price Service, Fuel Oil and Oil Heat, and oil companies.

1966 Jan.	1966 July	1967 Jan.	1967 July	1968 Jan.	1968 July	1969 Jan.	1969 July	1970 Jan.	1970 July	1971 Jan.	1971 July	1972 Jan.	1972 July	1973 Jan.
10.3	9.9	10.5	11.0	11.0	11.4	11.0	11.0	11.0	11.4	12.2	11.9	11.9	11.9	11.9
10.4	10.0	10.6	11.0	11.0	11.4	11.0	11.0	11.0	11.4	12.2	11.9	11.9	11.9	11.9
10.2	9.6	10.4	10.8	10.8	11.2	10.8	10.8	10.8	11.2	12.0	11.7	11.7	11.7	11.7
--	--	--	--	--	--	--	--	--	--	--	--	--	--	--
--	--	--	--	--	--	--	--	--	--	--	--	--	--	--
10.6	10.2	10.6	11.1	11.8	12.2	11.8	11.8	11.1	11.5	11.8	11.8	11.5	11.5	11.5
10.2	9.8	10.3	10.7	10.7	11.1	10.7[3/]	10.7[3/]	10.7[3/]	11.1[e/]	11.5[e/]	11.6	11.6	11.6	11.6
--	--	--	--	--	--	--	--	--	--	--	--	--	--	--
9.8	9.3	9.5	9.3	10.0	9.8	10.3	9.9	11.3	11.3	11.5	11.0	11.0	11.0	11.0
9.9	9.9	10.1	10.1	10.4	10.4	10.6	10.6	10.9	10.9	11.4	11.1	11.1	11.1	11.1
--	--	--	--	--	--	--	--	--	--	--	--	--	--	--
--	--	--	--	--	--	--	--	--	--	--	--	--	--	--
--	--	--	--	--	--	--	--	--	--	--	--	--	--	--
--	--	--	--	--	--	--	--	--	--	--	--	--	--	--
--	--	--	--	--	--	--	--	--	--	--	--	--	--	--
10.2	9.9	10.4	10.4	11.0	10.7	10.9	10.5	11.3	10.8	10.8	10.8	10.8	10.8	10.9
--	--	--	--	--	--	--	--	--	--	--	--	--	--	--
9.9	9.0	9.5	9.5	9.9	9.9	9.9	9.9	10.5	10.3	10.3	11.0	11.0	11.0	11.5
--	--	--	--	--	--	--	--	--	--	--	--	--	--	--
--	--	--	--	--	--	--	--	--	--	--	--	--	--	--
--	--	--	--	--	--	--	--	--	--	--	--	--	--	--
--	--	--	--	--	--	--	--	--	--	--	--	--	--	--
--	--	--	--	--	--	--	--	--	--	--	--	--	--	--
10.7	10.7	10.9	10.9	10.9	11.1	11.1	11.5	11.5	12.0[e/]	12.7	12.7	12.7	12.7	12.7

Table 8-d

WHOLESALE PRICE FOR NO. 6 FUEL OIL - FOB REFINERY OR TERMINAL 1/

(Dollars Per Barrel)

Census Region SMSA	1960		1961		1962		1963		1964		1965	
	Jan.	July	Jan.	July	Jan.	July	Jan.	July	Jan.	July	Jan.	July
New England												
Boston, Mass.	2.56	2.71	2.74	2.74	2.74	2.34	2.34	2.34	2.31	2.31	2.34	2.34
Middle Atlantic												
Albany, N.Y.	2.49	2.72	2.72	2.72	2.72	2.40	2.40	2.40	2.40	2.40	2.40	2.40
New York, N.Y.	2.42	2.65	2.65	2.65	2.65	2.28	2.28	2.28	2.28	2.28	2.28	2.28
Pittsburgh, Pa.	--	--	--	--	--	--	--	--	--	--	--	--
South Atlantic												
Atlanta, Ga.	--	--	--	--	--	--	--	--	--	--	--	--
Charlotte, N.C.	--	--	--	--	--	--	--	--	--	--	--	--
Washington, D.C.	2.63 e/	2.85 e/	2.85 e/	2.85 e/	2.85 e/	2.66 e/	2.66 e/	2.66 e/	2.66 e/	2.68 e/	2.68 e/	2.67 e/
Miami, Fla.	2.35	2.35	2.50	2.50	2.50	2.33	2.33	2.33	2.33	2.33	2.33	2.33
East North Central												
Chicago, Ill.	3.25 7/	3.21 7/	3.32 7/	3.21 7/	3.76 7/	3.76 7/	3.42 7/	3.42 7/	3.53 7/	3.21 2/	3.21 2/	3.11 2/
Detroit, Mich.	3.61	3.61	3.61	3.61	3.61	3.61	3.40	3.40	3.40	3.40	3.40	3.51 2/
Columbus, Ohio	--	--	--	--	--	--	--	--	--	--	--	--
East South Central												
Nashville, Tenn.	--	--	--	--	--	--	--	--	--	--	--	--
West South Central												
Houston, Tex.	--	--	--	--	--	--	--	--	--	--	--	--
New Orleans, La.	--	--	--	--	--	--	--	--	--	--	--	--
Tulsa, Okla.	--	--	--	--	--	--	--	--	--	--	--	--
West North Central												
Minneapolis-St. Paul, Minn.	3.53	3.36	3.36	3.47	3.47	3.47	3.47	3.47	3.47	3.36	3.57	3.57
Omaha, Neb.	--	--	--	--	--	--	--	--	--	--	--	--
St. Louis, Mo.	3.05	3.05	3.05	3.05	2.90	2.90	2.90	3.05	3.05	2.63	2.79	2.79
Mountain												
Denver, Colo.	--	--	--	--	--	--	--	--	--	--	--	--
Phoenix, Ariz.	--	--	--	--	--	--	--	--	--	--	--	--
Salt Lake City, Utah.	n.a.	n.a.	n.a.	n.a.	n.a.	n.a.	n.a.	3.32	3.32	3.10	3.20	3.00
Pacific												
Los Angeles, Calif.	--	--	--	--	--	--	--	--	--	--	--	--
San Francisco, Calif.	--	--	--	--	--	--	--	--	--	--	--	--
Seattle, Wash.	2.75 4/	2.75 4/	2.85 4/	2.88 4/	2.88 4/	2.98 4/	2.98 4/	2.88 4/	2.88 4/	2.88 4/	2.88 4/	2.88 4/

1/ Sulfur content is regular or no sulfur guarantee unless otherwise specified.
2/ Sulfur - 1% max.
3/ Sulfur - 0.5% max.
4/ Sulfur - 2% max.
5/ Sulfur - 0.3% max.
6/ Sulfur - 0.8% max.
7/ Low sulfur.
8/ Consumer price.
9/ Delivered price.

e/ Estimated.

n.a. - not available.

Source: Platt's Oilgram Price Service and oil companies.

1966		1967		1968		1969		1970		1971		1972		1973
Jan.	July	Jan.	July	Jan.	July	Jan.	July	Jan.	July	Jan.	July	Jan.	July	Jan.
2.34	2.34	2.31	2.31	2.34	2.34	2.34	1.92	2.05	2.65	4.15[2]	4.35[2]	4.91[3] [e]	4.43[3] [e]	4.93[3]
2.40	2.40	2.40[8]	2.40[8]	2.41	2.41	2.41	2.41	2.41	2.97	3.82	3.75	3.22	3.04	3.07
2.25	2.05	2.38[4]	2.38[4]	2.38[4]	2.38[4]	2.38[4]	2.38[4]	2.51[2]	3.28[2]	4.13[2]	4.33[2]	5.08[5]	4.53[5]	5.04[5]
--	--	--	--	--	--	--	--	--	--	--	--	--	--	--
--	--	--	--	--	--	--	--	--	--	--	--	--	--	--
--	--	--	--	--	--	--	--	--	--	--	--	--	--	--
2.67[e]	2.67[e]	2.67[e]	2.67[e]	2.67[e]	2.67[e]	2.73[2]	2.73[2]	2.76[2]	2.93[2]	3.03[2]	4.48[2]	4.68[2]	4.38[2]	4.68[6]
2.33	2.33	2.33	2.33	2.33	2.33	2.33	2.33	2.33	2.66	4.06	4.06	3.76	3.76	3.98
3.21[2]	3.05[2]	3.25[2]	3.17[2]	3.25[2]	3.36[2]	3.36[2]	3.25[2]	3.47[2]	3.89[2]	4.52[2]	4.62[2]	4.62[2]	4.62[2]	4.62[2]
3.51[2]	3.51[2]	3.51[2]	3.51[2]	3.51[2]	3.51[2]	3.51[2]	3.51[2]	3.51[2]	3.51[2]	4.31[2]	4.31[2]	4.73[2]	4.73[2]	4.73[2]
--	--	--	--	--	--	--	--	--	--	-- [9]	-- [9]	-- [9]	-- [9]	-- [9]
--	--	--	--	--	--	--	--	--	--	--	--	--	--	--
--	--	--	--	--	--	--	--	--	--	--	--	--	--	--
--	--	--	--	--	--	--	--	--	--	--	--	--	--	--
--	--	--	--	--	--	--	--	--	--	--	--	--	--	--
3.57	3.57	3.57	3.57	3.57	3.57	3.57	3.89	3.89	3.57	3.78	4.31	4.31	4.31	4.31
--	--	--	--	--	--	--	--	--	--	--	--	--	--	--
2.79	2.79	2.79	2.79	2.79	2.79	2.79	3.00	3.00	3.00	3.95	3.95	3.95	3.78	3.95
--	--	--	--	--	--	--	--	--	--	--	--	--	--	--
--	--	--	--	--	--	--	--	--	--	--	--	--	--	--
3.00	2.84	3.04	3.15	3.15	3.15	3.25	2.94	3.25	3.68	4.20	4.20	4.20	4.20	4.20
--	--	--	--	--	--	--	--	--	--	--	--	--	--	--
--	--	--	--	--	--	--	--	--	--	--	--	--	--	--
2.88[4]	2.88[4]	2.88[4]	2.88[4]	2.88[4]	2.88[4]	2.88[4]	2.88[4]	2.88[4]	3.11[4]	4.06[4]	4.36[4]	4.26[4]	4.26[4]	4.26[4]

Table 8-e

WHOLESALE PRICES* TO SERVICE STATION DEALERS OF MAJOR BRAND REGULAR GRADE GASOLINE

(¢/Gal.)

Census Region SMSA	1960		1961		1962		1963		1964		1965	
	Jan.	July	Jan.	July	Jan.	July	Jan.	July	Jan.	July	Jan.	July
New England												
Boston, Mass.	9.9	14.4	14.4	15.2	14.4	12.4	12.5	15.2	15.1	15.1	13.2	15.1
Middle Atlantic												
Albany, N.Y.	15.3	15.4	14.7[e/]	14.8[e/]	12.9	15.3	14.6	12.9	13.8	14.6	15.4	15.4
New York, N.Y.	16.2	16.2	16.2	16.2	16.2	16.2	16.2	16.2	16.2	16.2	16.2	16.2
Pittsburgh, Pa.	15.9	16.1	16.3	15.0	14.6	14.6	13.8	15.3	14.6	11.9	12.9	14.9
South Atlantic												
Atlanta, Ga.	15.9	14.4	15.2	15.2	15.2	14.4	16.9	16.2	16.2	16.2	16.2	16.2
Charlotte, N.C.	15.6	14.9	15.8	15.8	15.8	15.0	14.2	15.8	15.8	14.2	15.8	15.8
Washington, D.C.	14.9	14.9	15.9	14.9	15.9	15.4	14.1	15.9	15.2	14.9	14.9	14.9
Miami, Fla.	16.6	15.9	15.9	15.9	15.9	15.9	16.6	16.6	16.4	16.4	16.4	16.4
East North Central												
Chicago, Ill.	16.4	16.4	17.2	16.9	15.9	12.9	16.9	14.5	13.7	14.8	15.5	16.9
Detroit, Mich.	11.3	16.8	16.8	10.1	8.7	9.7	13.1	15.9	12.4	12.4	12.4	15.9
Columbus, Ohio	16.1	16.1	16.9	16.9	15.9	15.9	15.9	15.9	15.9	15.9	15.9	15.9
East South Central												
Nashville, Tenn.	14.9	14.9	14.9[e/]	14.9[e/]	14.9[e/]	14.9[e/]	14.9[e/]	14.9[e/]	15.7[e/]	15.7[e/]	15.7[e/]	15.7[e/]
West South Central												
Houston, Tex.	15.9	15.9	16.4	15.6	14.4	15.2	14.0	14.0	14.0	14.0	13.0	14.6
New Orleans, La.	13.9	10.9	13.9	10.9	13.9	13.9	13.9	13.9	13.9	13.9	13.9	13.9
Tulsa, Okla.	15.5	16.5	16.5	15.5	15.7	16.7	14.3	16.7	15.7	14.1	15.7	15.7
West North Central												
Minneapolis-St. Paul, Minn.	15.3	16.8	16.8	15.3	13.7	13.0	15.3	16.0	14.5	13.8	13.8	14.5
Omaha, Neb.	15.7	15.7	10.4	17.2	17.2	16.4	15.7	17.7	15.7	15.7	12.4	16.4
St. Louis, Mo.	11.9	15.8	16.4	9.9	15.4	6.4	15.9	15.9	11.4	14.2	15.4	15.4
Mountain												
Denver, Colo.	12.3	17.6	17.6	16.8	14.6	12.9	16.8	16.1	10.9	16.8	14.6	16.1
Phoenix, Ariz.	19.5	19.5	19.5	19.5	19.5	19.5	19.5	18.0	12.4	12.4	16.9	16.9
Salt Lake City, Utah	19.3	17.4	17.4	16.8	16.8	16.0	16.8	15.8	16.8	16.8	16.8	16.8
Pacific												
Los Angeles, Calif.	17.8	17.8	17.8	17.8	17.8	17.8	17.8	12.9	15.9	14.4	15.1	15.9
San Francisco, Calif.	18.3	18.3	18.3	18.3	18.3	18.3	18.3	16.8	15.9	15.9	15.9	15.9
Seattle, Wash.	19.0	19.0	19.0	19.0	19.0	19.0	17.0	17.5	14.7	14.7	12.4	15.4

e/ Estimate.

* Excludes taxes.

Note: Prices are as of first of month.

Source: Platt's Oilgram Price Service; various oil companies.

1966		1967		1968		1969		1970		1971		1972		1973
Jan.	July	Jan.	July	Jan.	July	Jan.	July	Jan.	July	Jan.	July	Jan.	July	Jan.
15.1	15.5	16.0	16.3	16.6	16.6	17.45	17.95	17.9	18.6	19.4	19.4	19.4	17.3	18.0
15.4	15.9	16.4	17.0	17.0	17.0	17.8	17.2	18.4	16.6	17.2	17.1	16.4	18.4	19.1
16.2	16.2	16.7	17.0	17.0	17.0	18.0	18.6	18.6	19.3	20.0	20.0	20.0	20.1	20.1
14.9	15.4	15.4	15.4	16.0	16.0	16.7	17.3	17.3	18.0	18.7	17.3	18.7	18.7	18.7
16.2	16.7	16.7	16.3	16.3	16.3	16.3	16.9	16.9	17.6	18.4	18.4	18.3	18.3	18.3
15.8	16.2	16.2	16.8	16.8	16.8	16.8	17.4	17.4	18.0	18.8	16.0	18.7	17.4	18.8
14.9	15.4	15.4	15.8	15.8	16.0	16.7	17.3	17.3	18.0	18.7	18.7	18.8	18.8	18.7
16.4	16.9	13.7	12.0	16.9	16.9	13.7	17.5	17.5	18.2	18.9	17.2	18.9	18.9	18.9
16.9	16.9	16.2	17.5	17.5	17.5	17.5	18.2	18.2	18.9	19.6	18.2	17.5	14.7	19.6
15.9	15.9	15.9	16.5	16.5	16.5	16.5	14.5	14.5	18.7	11.0	13.1	13.1	17.9	17.9
15.9	16.4	16.4	16.7	16.7	17.5	17.5	18.2	18.2	18.9	15.4	19.6	19.6	18.2	19.6
/15.7 [e/]	16.1 [e/]	16.1	16.5	16.5	17.1	17.8	17.8	18.5	18.5	19.2	15.0	17.8	17.8	17.8
14.6	15.1	15.1	15.1	15.1	15.7	15.7	12.5	16.3	17.0	17.8	17.7	17.7	16.3	16.3
13.9	14.4	14.9	14.9	14.9	15.5	15.5	16.1	16.1	16.8	17.5	17.5	14.7	16.1	16.2
15.7	15.7	16.2	16.8	16.8	16.8	16.8	17.5	17.5	18.2	18.9	18.9	18.9	18.9	18.9
16.4	16.9	16.9	16.2	17.5	16.9	16.9	18.2	16.8	18.9	17.5	18.2	19.6	18.2	18.2
16.4	16.9	16.9	17.5	17.5	16.2	17.5	18.2	18.2	18.9	16.1	19.6	13.3	18.2	18.2
15.9	16.4	16.4	17.3	17.3	17.3	17.5	18.0	18.2	18.9	19.6	19.6	19.6	18.2	19.6
13.7	16.1	16.9	17.0	17.5	15.5	17.5	15.4	12.6	18.9	18.2	17.5	17.5	18.2	16.8
15.4	16.9	17.4	17.4	17.4	17.4	17.4	18.0	18.0	18.7	19.5	19.5	19.5	18.1	18.8
16.8	14.2	12.5	16.4	16.4	16.4	16.4	17.1	17.1	17.8	18.5	18.5	18.5	18.5	18.5
12.9	15.9	14.2	16.4	16.4	16.4	16.4	17.0	17.0	17.7	18.4	18.4	16.9	15.6	17.9
15.9	15.9	16.4	16.4	16.4	16.4	16.4	17.0	17.0	17.7	18.4	18.4	18.4	17.0	17.9
13.9	12.4	15.2	14.8	14.8	15.9	15.9	16.5	16.5	17.2	17.9	17.9	16.8	15.5	17.9

Table 8-f
INDUSTRIAL WHOLESALE PRICES OF COAL[1/]
($/Ton)

Metropolitan Area	1960	1961	1962	1963	1964	1965
Albany	4.62	4.59	4.65	4.45	4.48	4.50
New York City	7.10	6.37	5.78	5.60	6.16	6.03
Pittsburgh, Pa.[2/]	5.29	5.17	5.07	4.90	5.07	5.07
Atlanta	4.75	4.54	4.57	4.68	4.92	5.00
Charlotte	4.75	4.54	4.57	4.68	4.92	5.00
Nashville	4.75	4.54	4.57	4.68	4.92	5.00
Columbus[3/]	3.85	3.77	3.72	3.70	3.69	3.71
Chicago	4.35	4.35	4.37	4.36	4.46	4.53
Detroit	4.62	4.47	4.48	4.53	4.45	4.57
St. Louis	3.95	3.95	3.95	3.95	4.00	4.00
Minneapolis-St. Paul[4/]	4.61	4.48	3.78	3.99	4.03	3.94
Omaha	4.31	4.28	4.16	4.16	4.08	4.15
Denver	5.55	4.24	4.17	4.11	4.12	4.14
Salt Lake City[5/]	3.45	3.39	3.20	3.18	3.15	3.11
Seattle[6/]	7.54	7.24	6.94	7.26	8.45	9.07

1/ Wholesale prices represent the FOB mine price.
2/ Pennsylvania average value per ton.
3/ Ohio average per ton.
4/ Pike County, Easton Kentucky, average value per ton.
5/ Wyoming average value per ton 1960-1968.
6/ Washington average value per ton.
7/ Decline from 1971 to 1972 due to large increase in proportion of low cost strip mine production.

Source: Coal producers and U.S. Bureau of Mines.

1966	1967	1968	1969	1970	1971	1972	1973
4.50	4.65	4.82	5.03	5.40	7.35	7.85	8.31
6.60	6.60	6.58	7.09	11.72	13.78	14.80	15.74
5.22	5.28	5.37	5.87	7.27	8.52	9.68	9.68
5.24	5.67	6.01	6.92	10.52	13.21	14.10	14.76
5.24	5.67	6.01	6.92	10.52	13.21	14.10	14.76
5.24	5.67	6.01	6.92	10.52	13.21	14.10	14.76
3.79	3.84	3.96	4.10	4.74	5.25	n.a.	n.a.
4.74	4.79	5.05	5.22	5.51	6.00	7.00	9.35
4.77	4.97	5.20	5.65	6.90	8.13	8.80	9.43
4.04	4.04	4.35	4.77	6.04	7.23	8.02	8.10
4.11	4.33	4.00	4.28	7.64	8.58	n.a.	n.a.
4.14	4.20	4.20	4.33	4.39	4.87	4.94	5.07
4.20	4.16	4.25	4.52	4.82	5.25	4.64	5.86
3.23	3.31	3.16	2.90	4.77	4.75	5.25	5.25
8.77	8.78	9.02	8.21	12.81	6.26[7/]	n.a.	n.a.

Table 9-a-1

BREAKDOWN OF THE PRICE OF ELECTRICITY INTO COST COMPONENTS
RESIDENTIAL NON-SUBSTITUTABLE BASE LOAD[2/]

Census Region	Cents Per Kwh									
	Production		Transpor-tation		Distribution		Taxes		Total	
SMSA	1960	1971	1960	1971	1960	1971	1960	1971	1960	1971
New England										
Boston	1.050	1.260	--	--	2.704	2.495	-0-	-0-	3.754	3.755
Middle Atlantic										
Albany	0.830	1.000	--	--	1.523	1.500	0.070	0.175	2.423	2.675
New York	1.290	1.620	--	--	2.735	1.910	0.080	0.298	4.105	3.828
South Atlantic										
Atlanta	0.680	0.830	--	--	1.327	1.120	0.060	0.058	2.067	2.008
Charlotte	0.580	0.750	--	--	1.854	1.628	0.072	0.072	2.506	2.450
Washington, D.C.	0.820	1.160	--	--	1.725	1.160	0.052	0.093	2.597	2.413
Miami	0.800	0.770	--	--	1.873	1.608	0.080	0.095	2.753	2.473
East North Central										
Chicago	0.930	1.040	--	--	1.948	1.903	0.161	0.237	3.039	3.180
Detroit	0.880	1.060	--	--	2.016	1.810	0.086	0.155	2.982	2.985
Columbus	0.880	0.930	--	--	2.058	1.628	0.088	0.102	3.026	2.660
East South Central										
Nashville	0.410	0.680	--	--	1.280	1.503	0.049	0.065	1.739	2.248
West South Central										
Houston	0.540	0.470	--	--	1.602	2.000	-0-	0.123	2.142	2.593
New Orleans	0.730	0.540	--	--	1.831	2.415	0.052	0.178	2.613	3.133
Tulsa	0.650	0.520	--	--	2.324	2.363	0.060	0.087	3.034	2.970
West North Central										
Minneapolis	0.860	1.040	--	--	2.130	1.828	0.044	0.172	3.034	3.040
Omaha - Public	0.770	0.730	--	--	1.497	1.545	-0-[1/]	0.080	2.267	2.355
- Private	0.990	0.950	--	--	1.771	2.130	0.055[1/]	0.093	2.816	3.173
St. Louis	0.790	1.040	--	--	1.875	1.548	0.055	0.165	2.720	2.753
Mountain										
Phoenix	0.810	0.640	--	--	1.826	2.040	0.039	0.080	2.675	2.760
Salt Lake City	0.830	0.880	--	--	1.510	1.535	0.060	0.108	2.400	2.523
Pacific										
Los Angeles - Public	0.660	0.640	--	--	1.368	1.418	0.060	0.185	2.088	2.243
- Private	0.740	0.870	--	--	1.525	1.498	0.067	0.167	2.332	2.535
San Francisco	0.810	0.830	--	--	1.452	1.200	0.068	0.183	2.330	2.213

1/ The publicly owned utility is located in a different city and state.
2/ Does not assume electric substitutable base load or electric space heating.

Source: Based on data from Forms 1 and 1-M Annual Reports of Electric Utilities to the FPC, the FPC National Electric Rate Book, and Commerce Clearing House.

Percent Distribution											
Production		Transportation		Distribution		Taxes		Total			
1960	1971	1960	1971	1960	1971	1960	1971	1960	1971		
28.0	33.6	--	--	72.0	66.4	-0-	-0-	100.0	100.0		
34.2	37.4	--	--	62.9	56.1	2.9	6.5	100.0	100.0		
31.5	42.3	--	--	66.6	49.9	1.9	7.8	100.0	100.0		
32.9	41.3	--	--	64.2	55.8	2.9	2.9	100.0	100.0		
23.1	30.6	--	--	74.0	66.5	2.9	2.9	100.0	100.0		
31.6	48.1	--	--	66.4	48.1	2.0	3.8	100.0	100.0		
29.1	31.1	--	--	68.0	65.0	2.9	3.9	100.0	100.0		
30.6	32.7	--	--	64.1	59.8	5.3	7.5	100.0	100.0		
29.5	35.5	--	--	67.6	60.6	2.9	3.9	100.0	100.0		
29.1	35.0	--	--	68.0	61.2	2.9	3.8	100.0	100.0		
23.6	30.2	--	--	73.6	66.9	2.8	2.9	100.0	100.0		
25.2	18.1	--	--	74.8	77.1	-0-	4.8	100.0	100.0		
27.9	17.2	--	--	70.1	77.1	2.0	5.7	100.0	100.0		
21.4	17.5	--	--	76.6	79.6	2.0	2.9	100.0	100.0		
28.3	34.2	--	--	70.2	60.1	1.5	5.7	100.0	100.0		
34.0	31.0	--	--	66.0	65.6	-0- 1/	3.4	100.0	100.0		
35.2	30.0	--	--	62.9	67.1	1.9 1/	2.9	100.0	100.0		
29.0	37.8	--	--	69.0	56.2	2.0	6.0	100.0	100 [illegible]		
30.3	23.2	--	--	68.3	73.9	1.4	2.9	100.0	100.0		
34.6	34.9	--	--	62.9	60.8	2.5	4.3	100.0	100.0		
31.6	28.6	--	--	65.5	63.2	2.9	8.2	100.0	100.0		
31.7	34.3	--	--	65.4	59.1	2.9	6.6	100.0	100.0		
34.8	37.5	--	--	62.3	54.2	2.9	8.3	100.0	100.0		

Table 9-a-2

BREAKDOWN OF THE PRICE OF ELECTRICITY INTO COST COMPONENTS RESIDENTIAL SUBSTITUTABLE BASE LOAD AND SPACE HEATING[1/]

Census Region SMSA	Cents Per Kwh									
	Production		Transportation		Distribution		Taxes		Total	
	1960	1971	1960	1971	1960	1971	1960	1971	1960	1971
New England										
Boston	1.050	1.260	--	--	1.998	0.748	-0-	-0-	3.048	2.008
Middle Atlantic										
Albany	0.830	1.000	--	--	0.599	0.550	0.042	0.108	1.471	1.658
South Atlantic										
Atlanta	0.680	0.830	--	--	0.583	0.463	0.038	0.039	1.301	1.332
Charlotte	0.580	0.750	--	--	0.918	0.376	0.045	0.034	1.543	1.160
Washington, D.C.	0.820	1.160	--	--	0.984	0.074	0.036	0.049	1.840	1.283
Miami	0.800	0.770	--	--	0.887	0.869	0.030	0.040	1.717	1.679
East North Central										
Chicago	0.930	1.040	--	--	0.857	0.245	0.100	0.205	1.887	1.490
Detroit	0.880	1.060	--	--	1.142	0.467	0.060	0.061	2.082	1.588
Columbus	0.880	0.930	--	--	0.685	0.361	0.047	0.051	1.612	1.342
East South Central										
Nashville	0.410	0.680	--	--	0.277	0.247	0.020	0.027	0.707	0.954
West South Central										
Houston	0.540	0.470	--	--	0.610	0.532	-0-	0.050	1.150	1.052
New Orleans	0.730	0.540	--	--	0.791	0.709	0.031	0.075	1.552	1.324
Tulsa	0.650	0.520	--	--	0.532	0.506	0.020	0.031	1.022	1.057
West North Central										
Minneapolis	0.860	1.040	--	--	0.900	0.354	0.027	0.085	1.787	1.479
Omaha - Public	0.770	0.730	--	--	0.702	0.179	-0-[2/]	0.031	1.472	0.940
- Private	0.990	0.950	--	--	0.742	0.500	0.035[2/]	0.043	1.767	1.493
Mountain										
Phoenix	0.810	0.640	--	--	1.020	0.470	0.028	0.034	1.858	1.144
Salt Lake City	0.830	0.880	--	--	0.715	0.217	0.039	0.049	1.584	1.146
Pacific										
Los Angeles - Public	0.660	0.640	--	--	0.632	0.549	0.039	0.107	1.331	1.296
- Private	0.740	0.870	--	--	0.616	0.379	0.041	0.125	1.397	1.374
San Francisco	0.810	0.830	--	--	0.469	0.449	0.038	0.115	1.317	1.394

1/ Assumes electric substitutable base load and space heating.
2/ The publicly owned utility is located in a different city and state.

Source: Based on data from Forms 1 and 1-M Annual Reports of Electric Utilities to the FPC, the FPC National Electic Rate Book, and Commerce Clearing House.

Percent Distribution									
Production		Transportation		Distribution		Taxes		Total	
1960	1971	1960	1971	1960	1971	1960	1971	1960	1971
34.4	62.7	--	--	65.6	37.3	-0-	-0-	100.0	100.0
56.4	60.3	--	--	40.7	33.2	2.9	6.5	100.0	100.0
52.3	62.3	--	--	44.8	34.8	2.9	2.9	100.0	100.0
37.6	64.7	--	--	59.5	32.4	2.9	2.9	100.0	100.0
44.5	90.4	--	--	53.5	5.8	2.0	3.8	100.0	100.0
46.6	45.9	--	--	51.7	51.7	1.7	2.4	100.0	100.0
49.3	69.8	--	--	45.4	16.4	5.3	13.8	100.0	100.0
42.3	66.8	--	--	54.8	29.4	2.9	3.8	100.0	100.0
54.6	69.3	--	--	42.5	26.9	2.9	3.8	100.0	100.0
58.0	71.3	--	--	39.2	25.9	2.8	2.8	100.0	100.0
47.0	44.7	--	--	53.0	50.5	-0-	4.8	100.0	100.0
47.0	40.8	--	--	51.0	53.5	2.0	5.7	100.0	100.0
63.6	49.2	--	--	34.4	47.9	2.0	2.9	100.0	100.0
48.1	70.3	--	--	50.4	23.9	1.5	5.8	100.0	100.0
52.3	77.7	--	--	47.7	19.0	-0-[2/]	3.3	100.0	100.0
56.0	63.6	--	--	42.0	33.5	2.0[2/]	2.9	100.0	100.0
43.6	55.9	--	--	54.9	41.1	1.5	3.0	100.0	100.0
52.4	76.8	--	--	45.1	18.9	2.5	4.3	100.0	100.0
49.6	49.4	--	--	47.5	42.4	2.9	8.2	100.0	100.0
53.0	63.3	--	--	44.1	27.6	2.9	9.1	100.0	100.0
61.5	59.5	--	--	35.6	32.2	2.9	8.3	100.0	100.0

Table 9-a-3

BREAKDOWN OF THE PRICE OF ELECTRICITY INTO COST COMPONENTS
SMALL COMMERCIAL

Census Region SMSA	Cents Per Kwh Production 1960	Production 1971	Transportation 1960	Transportation 1971	Distribution 1960	Distribution 1971	Taxes 1960	Taxes 1971	Total 1960	Total 1971
New England										
Boston	1.050	1.260	--	--	3.239	3.179	-0-	-0-	4.289	4.439
Middle Atlantic										
Albany	0.830	1.000	--	--	1.874	1.750	0.081	0.193	2.785	2.943
New York	1.290	1.620	--	--	2.052	2.656	0.102	0.371	3.444	4.647
South Atlantic										
Atlanta	0.680	0.830	--	--	2.200	2.123	0.086	0.089	2.966	3.042
Charlotte	0.580	0.750	--	--	1.820	1.652	0.072	0.072	2.472	2.474
Washington, D.C.	0.820	1.160	--	--	1.880	1.490	0.054	0.106	2.754	2.756
Miami	0.800	0.770	--	--	3.568	2.648	0.131	0.137	4.499	3.555
East North Central										
Chicago	0.930	1.040	--	--	2.770	2.607	0.207	0.293	3.907	3.940
Detroit	0.880	1.060	--	--	2.470	2.083	0.101	0.126	3.451	3.269
Columbus	0.880	0.930	--	--	2.190	1.545	0.092	0.099	3.162	2.574
East South Central										
Nashville	0.410	0.680	--	--	0.890	1.125	0.039	0.054	1.339	1.859
West South Central										
Houston	0.540	0.470	--	--	1.830	1.844	-0-	0.116	2.370	2.430
New Orleans	0.730	0.540	--	--	2.913	2.718	0.073	0.195	3.716	3.453
Tulsa	0.650	0.520	--	--	2.500	2.573	0.063	0.093	3.213	3.186
West North Central										
Minneapolis	0.860	1.040	--	--	2.970	2.085	0.058	0.191	3.888	3.316
Omaha - Public	0.770	0.730	--	--	2.056	1.645	-0- 1/	0.083	2.826	2.458
- Private	0.990	0.950	--	--	2.015	2.248	0.060 1/	0.096	3.065	3.294
St. Louis	0.790	1.040	--	--	2.557	2.340	0.067	0.217	3.414	3.595
Mountain										
Phoenix	0.810	0.640	--	--	1.919	2.142	0.041	0.197	2.770	2.979
Salt Lake City	0.830	0.880	--	--	1.741	1.864	0.064	0.123	2.635	2.867
Pacific										
Los Angeles - Public	0.660	0.640	--	--	1.555	1.767	0.066	0.217	2.281	2.624
- Private	0.740	0.870	--	--	2.180	2.029	0.088	0.206	3.008	3.105
San Francisco	0.810	0.830	--	--	2.192	2.395	0.090	0.290	3.092	3.515

1/ Publicly owned utility is in a different city and state from the privately owned utility.

Source: Based on data from Forms 1 and 1-M Annual Reports of Electric Utilities to the FPC, the FPC National Electric Rate Book, and Commerce Clearing House.

Percent Distribution									
Production		Transportation		Distribution		Taxes		Total	
1960	1971	1960	1971	1960	1971	1960	1971	1960	1971
24.5	28.4	--	--	75.5	71.6	-0-	-0-	100.0	100.0
29.8	34.0	--	--	67.3	59.5	2.9	6.5	100.0	100.0
37.4	34.9	--	--	59.6	57.1	3.0	8.0	100.0	100.0
22.9	27.3	--	--	74.2	69.8	2.9	2.9	100.0	100.0
23.5	30.3	--	--	73.6	66.8	2.9	2.9	100.0	100.0
29.8	42.1	--	--	68.3	54.1	1.9	3.8	100.0	100.0
17.8	21.7	--	--	79.3	74.5	2.9	3.8	100.0	100.0
23.8	26.4	--	--	70.9	66.2	5.3	7.4	100.0	100.0
25.5	32.4	--	--	71.6	63.7	2.9	3.9	100.0	100.0
27.8	36.1	--	--	69.3	60.0	2.9	3.9	100.0	100.0
30.6	36.6	--	--	66.5	60.5	2.9	2.9	100.0	100.0
22.8	19.3	--	--	77.2	75.9	-0-	4.8	100.0	100.0
19.6	15.6	--	--	78.4	78.7	2.0	5.7	100.0	100.0
20.2	16.3	--	--	77.8	80.8	2.0	2.9	100.0	100.0
22.1	31.4	--	--	76.4	62.9	1.5	5.7	100.0	100.0
27.2	29.7	--	--	72.8	66.9	-0-[1/]	3.4	100.0	100.0
32.3	28.9	--	--	65.7	68.2	2.0[1/]	2.9	100.0	100.0
23.1	28.9	--	--	74.9	65.1	2.0	6.0	100.0	100.0
29.2	21.5	--	--	69.3	71.9	1.5	6.6	100.0	100.0
31.5	30.7	--	--	66.1	65.0	2.4	4.3	100.0	100.0
28.9	24.4	--	--	68.2	67.3	2.9	8.3	100.0	100.0
24.6	28.0	--	--	72.5	65.4	2.9	6.6	100.0	100.0
26.2	23.6	--	--	70.9	68.1	2.9	8.3	100.0	100.0

Table 9-a-4

BREAKDOWN OF THE PRICE OF ELECTRICITY INTO COST COMPONENTS
LARGE COMMERCIAL

Census Region SMSA	Cents Per Kwh									
	Production		Transportation		Distribution		Taxes		Total	
	1960	1971	1960	1971	1960	1971	1960	1971	1960	1971
New England										
Boston	1.050	1.260	--	--	1.007	1.002	-0-	-0-	2.057	2.262
Middle Atlantic										
Albany	0.830	1.000	--	--	0.324	0.245	0.035	0.087	1.189	1.332
New York	1.290	1.620	--	--	0.595	1.303	0.057	0.246	1.942	3.169
South Atlantic										
Atlanta	0.680	0.830	--	--	0.775	0.709	0.044	0.046	1.499	1.585
Charlotte	0.580	0.750	--	--	0.500	0.366	0.032	0.033	1.112	1.149
Washington, D.C.	0.820	1.160	--	--	0.753	0.476	0.031	0.065	1.604	1.701
Miami	0.800	0.770	--	--	0.831	0.850	0.049	0.065	1.680	1.685
East North Central										
Chicago	0.930	1.040	--	--	0.561	0.691	0.083	0.138	1.574	1.869
Detroit	0.880	1.060	--	--	1.103	0.897	0.059	0.078	2.042	2.035
Columbus	0.880	0.930	--	--	0.716	0.622	0.048	0.062	1.644	1.614
East South Central										
Nashville	0.410	0.680	--	--	0.295	0.391	0.021	0.036	0.726	1.103
West South Central										
Houston	0.540	0.470	--	--	0.560	0.683	-0-	0.058	1.100	1.211
New Orleans	0.730	0.540	--	--	0.588	0.704	0.026	0.075	1.344	1.319
Tulsa	0.650	0.520	--	--	0.519	0.746	0.023	0.038	1.192	1.304
West North Central										
Minneapolis	0.860	1.040	--	--	0.891	0.644	0.026	0.103	1.777	1.787
Omaha - Public	0.770	0.730	--	--	0.320	0.501	-0- 1/	0.043	1.090	1.274
- Private	0.990	0.950	--	--	0.741	0.724	0.035 1/	0.050	1.766	1.724
St. Louis	0.790	1.040	--	--	0.665	0.475	0.029	0.097	1.484	1.612
Mountain										
Phoenix	0.810	0.640	--	--	0.597	0.820	0.021	0.104	1.428	1.564
Salt Lake City	0.830	0.880	--	--	0.481	0.474	0.033	0.061	1.344	1.415
Pacific										
Los Angeles - Public	0.660	0.640	--	--	0.297	0.155	0.029	0.072	0.986	0.867
- Private	0.740	0.870	--	--	0.437	0.315	0.035	0.084	1.212	1.269
San Francisco	0.810	0.830	--	--	0.489	0.551	0.039	0.124	1.338	1.505

1/ Publicly owned utility is in a separate city and state from the privately owned utility.

Source: Based on data from Forms 1 and 1-M Annual Reports of Electric Utilities to the FPC, the FPC National Electric Rate Book, and Commerce Clearing House.

Percent Distribution									
Production		Transportation		Distribution		Taxes		Total	
1960	1971	1960	1971	1960	1971	1960	1971	1960	1971
51.0	55.7	--	--	49.0	44.3	-0-	-0-	100.0	100.0
69.8	75.1	--	--	27.3	18.4	2.9	6.5	100.0	100.0
66.4	51.1	--	--	30.7	41.1	2.9	7.8	100.0	100.0
45.4	52.4	--	--	51.7	44.7	2.9	2.9	100.0	100.0
52.1	65.3	--	--	45.0	31.8	2.9	2.9	100.0	100.0
51.1	68.2	--	--	47.0	28.0	1.9	3.8	100.0	100.0
47.6	45.7	--	--	49.5	50.4	2.9	3.9	100.0	100.0
59.1	55.6	--	--	35.6	37.0	5.3	7.4	100.0	100.0
43.1	52.1	--	--	54.0	44.1	2.9	3.8	100.0	100.0
53.5	57.6	--	--	43.6	38.5	2.9	3.9	100.0	100.0
56.5	61.7	--	--	40.6	35.4	2.9	2.9	100.0	100.0
49.1	38.8	--	--	50.9	56.4	-0-	4.8	100.0	100.0
54.3	40.9	--	--	43.8	53.4	1.9	5.7	100.0	100.0
54.5	39.9	--	--	43.6	57.2	1.9	2.9	100.0	100.0
48.4	58.2	--	--	50.1	36.0	1.5	5.8	100.0	100.0
70.6	57.3	--	--	29.4	39.3	-0-[1/]	3.4	100.0	100.0
56.1	55.1	--	--	41.9	42.0	2.0[1/]	2.9	100.0	100.0
53.2	64.5	--	--	44.8	29.5	2.0	6.0	100.0	100.0
56.7	40.9	--	--	41.8	52.5	1.5	6.6	100.0	100.0
61.8	62.2	--	--	35.8	33.5	2.4	4.3	100.0	100.0
67.0	73.8	--	--	30.1	17.9	2.9	8.3	100.0	100.0
61.1	68.6	--	--	36.0	24.8	2.9	6.6	100.0	100.0
60.5	55.2	--	--	36.6	36.6	2.9	8.2	100.0	100.0

Table 9-a-5

BREAKDOWN OF THE PRICE OF ELECTRICITY INTO COST COMPONENTS
INDUSTRIAL

Census Region	Censt Per Kwh									
	Production		Transportation		Distribution		Taxes		Total	
SMSA	1960	1971	1960	1971	1960	1971	1960	1971	1960	1971
New England										
Boston	1.050	1.260	--	--	0.154	0.247	-0-	-0-	1.204	1.507
Middle Atlantic										
New York	1.290	1.620	--	--	0.150	0.448	0.014	0.174	1.454	2.242
South Atlantic										
Atlanta	0.680	0.830	--	--	0.142	0.029	-0-	-0-	0.822	0.859
Miami	0.800	0.770	--	--	0.346	0.323	-0-	-0-	1.146	1.093
East North Central										
Chicago	0.930	1.040	--	--	0.005	0.132	0.052	0.093	0.987	1.265
Detroit	0.880	1.060	--	--	0.293	0.046	-0-	-0-	1.173	1.106
Columbus	0.880	0.930	--	--	0.189	0.155	-0-	-0-	1.069	1.085
East South Central										
Nashville	0.410	0.680	--	--	0.080	0.085	-0-	-0-	0.490	0.765
West South Central										
Houston	0.540	0.470	--	--	0.145	0.147	-0-	-0-	0.685	0.617
New Orleans	0.730	0.540	--	--	0.065	0.186	-0-	-0-	0.795	0.726
Tulsa	0.650	0.520	--	--	0.189	0.202	0.017	0.022	0.856	0.744
West North Central										
Minneapolis	0.860	1.040	--	--	0.106	0.042	0.014	0.032	0.980	1.114
Omaha - Public	0.770	0.730	--	--	0.032	0.065	-0-	-0-	0.802	0.895
- Private	0.990	0.950	--	--	n.a.	0.048	n.a.	-0-	n.a.	0.998
Mountain										
Phoenix	0.810	0.640	--	--	0.132	0.355	-0-	0.046	0.942	1.041
Salt Lake City	0.830	0.880	--	--	0.038	0.005	-0-	-0-	0.868	0.885
Pacific										
Los Angeles - Public	0.660	0.640	--	--	0.080	0.081	-0-	-0-	0.740	0.721
San Francisco	0.810	0.830	--	--	0.039	0.077	-0-	-0-	0.849	0.907

Note: Some of the cities which were included for other market sectors have not been included on the industrial table. It was felt that the estimates of the cost of generation were too crude to render a breakdown precise enough to be meaningful as prices paid by this category of users were so close to the wholesale level.

Source: Based on data from Forms 1 and 1-M Annual Reports of Electric Utilities to the FPC, the FPC National Electric Rate Book, and Commerce Clearing House.

Percent Distribution									
Production		Transportation		Distribution		Taxes		Total	
1960	1971	1960	1971	1960	1971	1960	1971	1960	1971
87.2	83.6	--	--	12.8	16.4	-0-	-0-	100.0	100.0
88.7	72.2	--	--	10.3	20.0	1.0	7.8	100.0	100.0
82.7	96.6	--	--	17.3	3.4	-0-	-0-	100.0	100.0
69.8	70.4	--	--	30.2	29.6	-0-	-0-	100.0	100.0
94.2	82.2	--	--	0.5	10.4	5.3	7.4	100.0	100.0
75.0	95.8	--	--	25.0	4.2	-0-	-0-	100.0	100.0
82.3	85.7	--	--	17.7	14.3	-0-	-0-	100.0	100.0
83.7	88.9	--	--	16.3	11.1	-0-	-0-	100.0	100.0
78.8	76.2	--	--	21.2	23.8	-0-	-0-	100.0	100.0
91.8	74.4	--	--	8.2	25.6	-0-	-0-	100.0	100.0
75.9	69.9	--	--	22.1	27.1	2.0	3.0	100.0	100.0
87.8	93.3	--	--	10.8	3.8	1.4	2.9	100.0	100.0
96.0	81.6	--	--	4.0	18.4	-0-	-0-	100.0	100.0
n.a.	95.2	--	--	n.a.	4.8	n.a.	-0-	100.0	100.0
86.0	61.5	--	--	14.0	34.1	-0-	4.4	100.0	100.0
95.6	99.4	--	--	4.4	0.6	-0-	-0-	100.0	100.0
89.2	88.8	--	--	10.8	11.2	-0-	-0-	100.0	100.0
95.4	91.5	--	--	4.6	8.5	-0-	-0-	100.0	100.0

Table 9-b-1

BREAKDOWN OF THE PRICE OF #2 FUEL OIL BY COST COMPONENTS
RESIDENTIAL[1]

Census Region SMSA	Production		Transportation		Distribution		Taxes		Total	
	1960	1971	1960	1971	1960	1971	1960	1971	1960	1971
	Cents Per Gallon									
New England										
Boston, Mass.	8.35	9.88	0.99	1.60	5.56	8.42	--	--	14.90	19.90
Middle Atlantic										
New York, N.Y.	8.35	9.88	0.99	0.73	5.10	8.22	0.44	1.18	14.88	20.00
South Atlantic										
Charlotte, N.C.	8.35	9.88	0.94	0.58	4.91	5.94	0.43	0.66	14.63	17.06
Washington, D.C.	8.35	9.88	1.02	0.67	4.92	8.31	0.29	0.78	14.58	19.64
East North Central										
Chicago, Ill.	8.35	9.88	0.88	0.88	4.93	6.43	0.44	0.90	14.60	18.09
Detroit, Mich.	8.35	9.88	1.58	1.58	4.96	6.35	0 46	0.74	15.35	18.55
West North Central										
Minneapolis-St. Paul, Minn.	8.62	9.75	1.62	1.62	4.69	.601	--	0.54	14.93	17.92
St. Louis, Mo.	8.62	9.75	0.48	0.43	5.80	8.06	0.30	0.76	15.20	19.00
Pacific										
Seattle, Wash.	9.40	11.15	0.16	0.16	7.26	9.62	0.71	1.10	17.53	22.03

	Percent Distribution									
New England										
Boston, Mass.	56.1	49.7	6.6	8.0	37.3	42.3	--	--	100.0	100.0
Middle Atlantic										
New York, N.Y.	56.1	49.4	6.7	3.6	34.3	41.1	2.9	5.9	100.0	100.0
South Atlantic										
Charlotte, N.C.	57.1	57.9	6.4	3.4	33.6	34.8	2.9	2.9	100.0	100.0
Washington, D.C.	57.3	50.3	7.0	3.4	33.7	42.3	2.0	4.0	100.0	100.0
East North Central										
Chicago, Ill.	57.2	54.4	10.9	9.1	31.4	33.5	--	3.0	100.0	100.0
Detroit, Mich.	54.4	53.3	10.3	8.5	32.3	34.2	3.0	4.0	100.0	100.0
West North Central										
Minneapolis-St. Paul, Minn.	57.7	54.4	10.9	9.1	31.4	33.5	--	3.0	100.0	100.0
St. Louis, Mo.	56.7	51.3	3.1	2.3	38.2	42.4	2.0	4.0	100.0	100.0
Pacific										
Seattle, Wash.	53.6	50.6	0.9	0.7	41.4	43.7	4.0	5.0	100.0	100.0

Note: All data are averages of the middle of January and July.
1/ Space heating.
Source: Based on data from Platt's Oilgram Price Service, Fuel Oil and Oil Heat, U.S. Bureau of Labor Statistics, Retail Prices and Indexes of Fuels and Electricity, Interstate Commerce Commission, Oil Companies, Bulk Carriers Conference, Inc., American Petroleum Institute and Commerce Clearing House.

Table 9-b-2

BREAKDOWN OF THE PRICE OF #2 FUEL OIL INTO COST COMPONENTS
SMALL COMMERCIAL

Census Region	Production		Transportation		Distribution		Taxes		Total	
SMSA	1960	1971	1960	1971	1960	1971	1960	1971	1960	1971
	Cents Per Gallon									
New England										
Boston, Mass.	8.35	9.88	0.99	1.60	5.06	7.92	--	--	14.40	19.40
Middle Atlantic										
New York, N.Y.	8.35	9.88	0.99	0.73	4.60	7.72	0.44	1.18	14.38	19.50
South Atlantic										
Charlotte, N.C.	8.35	9.88	0.94	0.58	4.41	5.44	0.43	0.66	14.13	16.56
Washington, D.C.	8.35	9.88	1.02	0.67	4.42	7.81	0.29	0.78	14.08	19.14
East North Central										
Chicago, Ill.	8.35	9.88	0.88	0.88	4.43	5.93	0.44	0.90	14.10	17.59
Detroit, Mich.	8.35	9.88	1.58	1.58	4.46	5.85	0.46	0.74	14.85	18.05
West North Central										
Minneapolis-St. Paul, Minn.	8.62	9.75	1.62	1.62	4.19	5.51	--	0.54	14.43	17.42
St. Louis, Mo.	8.62	9.75	0.48	0.43	5.30	7.56	0.30	0.76	14.70	18.50
Pacific										
Seattle, Wash.	9.40	11.15	0.16	0.16	6.76	9.12	0.71	1.10	17.03	21.53

				Percent Distribution						
New England										
Boston, Mass.	58.0	50.9	6.9	8.3	35.1	40.8	--	--	100.0	100.0
Middle Atlantic										
New York, N.Y.	58.1	50.7	6.9	3.7	32.0	39.6	3.0	6.0	100.0	100.0
South Atlantic										
Charlotte, N.C.	59.1	59.7	6.7	3.5	31.2	32.8	3.0	4.0	100.0	100.0
Washington, D.C.	59.3	51.6	7.3	3.5	31.4	40.8	2.0	4.1	100.0	100.0
East North Central										
Chicago, Ill.	59.2	56.2	6.3	5.0	31.4	33.7	3.1	5.1	100.0	100.0
Detroit, Mich.	56.2	54.7	10.6	8.8	30.0	32.4	3.1	4.1	100.0	100.0
West North Central										
Minneapolis-St. Paul, Minn.	59.8	56.0	11.2	9.3	29.0	31.6	--	3.1	100.0	100.0
St. Louis, Mo.	58.6	52.7	3.3	2.3	36.1	40.9	2.0	4.1	100.0	100.0
Pacific										
Seattle, Wash.	55.2	51.8	1.0	0.7	39.7	42.4	4.1	5.1	100.0	100.0

Note: All data are averages based on estimates for the middle of January and July.

Source: Platt's Oilgram Price Service, oil companies, American Petroleum Institute and Commerce Clearing House.

Table 9-b-3

BREAKDOWN OF THE PRICE OF #2 FUEL OIL INTO COST COMPONENTS
LARGE COMMERCIAL

Census Region	Production		Transportation		Distribution		Taxes		Total	
SMSA	1960	1971	1960	1971	1960	1971	1960	1971	1960	1971
					Cents Per Gallon					
New England										
Boston, Mass.	8.35	9.88	0.99	1.60	0.66	1.17	--	--	10.00	12.65
Middle Atlantic										
New York, N.Y.	8.35	9.88	0.99	0.73	0.91	1.84	0.30	0.75	10.55	13.20
South Atlantic										
Charlotte, N.C.	8.35	9.88	0.94	0.58	1.41	1.94	0.30	0.50	11.00	12.90
Washington, D.C.	8.35	9.88	1.02	0.67	1.33	1.60	0.20	0.50	10.90	12.65
East North Central										
Chicago, Ill.	8.35	9.88	0.88	0.88	0.57	1.09	0.30	0.60	10.10	12.45
Detroit, Mich.	8.35	9.88	1.58	1.58	1.02	0.34	0.30	0.50	11.25	12.30
West North Central										
Minneapolis-St. Paul, Minn.	8.62	9.75	1.62	1.62	0.31	0.48	--	0.35	10.55	12.20
St. Louis, Mo.	8.62	9.75	0.48	0.43	0.75	1.02	0.20	0.45	10.05	11.65
Pacific										
Seattle, Wash.	9.40	11.15	0.16	0.16	1.54	2.04	0.40	0.70	11.50	14.05

	Percent Distribution									
New England										
Boston, Mass.	83.5	78.1	9.9	12.6	6.6	9.3	--	--	100.0	100.0
Middle Atlantic										
New York, N.Y.	79.2	74.9	9.4	5.5	8.6	13.9	2.8	5.7	100.0	100.0
South Atlantic										
Charlotte, N.C.	75.9	76.6	8.6	4.5	12.8	15.0	2.7	3.9	100.0	100.0
Washington, D.C.	76.6	78.1	9.4	5.3	12.2	12.6	1.8	4.0	100.0	100.0
East North Central										
Chicago, Ill.	82.7	79.4	8.7	7.0	5.6	8.8	3.0	4.8	100.0	100.0
Detroit, Mich.	74.2	80.3	14.0	12.8	9.1	2.7	2.7	4.0	100.0	100.0
West North Central										
Minneapolis-St. Paul, Minn.	81.7	79.9	15.4	13.3	2.9	3.9	--	2.9	100.0	100.0
St. Louis, Mo.	85.8	83.7	4.8	3.7	7.4	8.7	2.0	3.9	100.0	100.0
Pacific										
Seattle, Wash.	81.7	79.4	1.4	1.1	13.4	14.5	3.5	5.0	100.0	100.0

Note: All data are averages of the middle of January and July.

Source: Based on data from Platt's Oilgram Price Service, oil companies, Interstate Commerce Commission, Bulk Carrier Conference, Inc., American Petroleum Institute, and Commerce Clearing House.

Table 9-b-4

BREAKDOWN OF THE PRICE OF #2 FUEL OIL BY COST COMPONENTS
INDUSTRIAL

Census Region SMSA	Production 1960	Production 1971	Transportation 1960	Transportation 1971	Distribution 1960	Distribution 1971	Taxes 1960	Taxes 1971	Total 1960	Total 1971
	Cents Per Gallon									
New England										
Boston, Mass.	8.35	9.88	0.99	1.60	0.66	1.17	--	--	10.00	12.65
Middle Atlantic										
New York, N.Y.	8.35	9.88	0.99	0.73	0.91	2.04	--	--	10.25	12.65
South Atlantic										
Charlotte, N.C.	8.35	9.88	0.94	0.58	1.91	1.94	0.10	0.30	11.30	12.70
Washington, D.C.	8.35	9.88	1.02	0.67	1.33	1.60	--	--	10.70	12.15
East North Central										
Chicago, Ill.	8.35	9.88	0.88	0.88	0.57	1.09	0.30	0.60	10.10	12.45
Detroit, Mich.	8.35	9.88	1.58	1.58	1.02	0.34	--	--	10.95	11.80
West North Central										
Minneapolis-St. Paul, Minn.	8.62	9.75	1.62	1.62	0.31	0.48	--	--	10.55	11.85
St. Louis, Mo.	8.62	9.75	0.48	0.43	0.75	1.02	--	--	9.85	11.20
Pacific										
Seattle, Wash.	9.40	11.15	0.16	0.16	1.54	2.04	0.40	0.70	11.50	14.05

	Percent Distribution									
New England										
Boston, Mass.	83.5	78.1	9.9	12.6	6.6	9.3	--	--	100.0	100.0
Middle Atlantic										
New York, N.Y.	81.4	78.1	9.7	5.8	8.9	16.1	--	--	100.0	100.0
South Atlantic										
Charlotte, N.C.	73.9	77.8	8.3	4.5	16.9	15.3	0.9	2.4	100.0	100.0
Washington, D.C.	78.0	81.3	9.5	5.5	12.4	13.2	--	--	100.0	100.0
East North Central										
Chicago, Ill.	82.7	79.4	8.7	7.0	5.6	8.8	3.0	4.8	100.0	100.0
Detroit, Mich.	76.3	83.7	14.4	13.4	9.3	2.9	--	--	100.0	100.0
West North Central										
Minneapolis-St. Paul, Minn.	81.7	82.3	15.4	13.7	2.9	4.0	--	--	100.0	100.0
St. Louis, Mo.	87.5	87.1	4.9	3.8	7.6	9.1	--	--	100.0	100.0
Pacific										
Seattle, Wash.	81.7	79.4	1.4	1.1	13.4	14.5	3.5	5.0	100.0	100.0

Note: All data are averages of the middle of January and July.

Source: Based on data from Platt's Oilgram Price Service, oil companies, Interstate Commerce Commission, Bulk Carrier Conference, Inc., American Petroleum Institute, and Commerce Clearing House.

Table 9-c-1

BREAKDOWN OF THE PRICE OF #6 FUEL OIL INTO COST COMPONENTS
LARGE COMMERCIAL

Census Region SMSA	Production		Transportation		Distribution		Taxes		Total	
	1960	1971	1960	1971	1960	1971	1960	1971	1960	1971
	Dollars Per Barrel									
New England										
Boston, Mass.	2.26	4.11	--	--	0.66	0.49	--	--	2.92	4.60
Middle Atlantic										
New York, N.Y.	2.25	3.50	--	--	0.63	1.11	0.09	0.28	2.97	4.89
South Atlantic										
Washington, D.C.	2.25	3.50	--	--	0.83	1.46	0.06	0.20	3.14	5.16
East North Central										
Chicago, Ill.	2.85	3.88	--	--	0.59	1.02	0.10	0.25	3.54	5.15
Detroit, Mich.	3.21	3.62	--	--	0.61	0.69	0.11	0.17	3.93	4.48
West North Central										
Minneapolis-St. Paul, Minn.	2.98	3.62	--	--	0.59	0.94	--	0.14	3.57	4.70
St. Louis, Mo.	2.71	3.32	--	--	0.55	0.88	0.07	0.17	3.33	4.37
Pacific										
Seattle, Wash.	2.44	3.66	--	--	0.56	1.00	0.12	0.23	3.12	4.89

					Percent Distribution					
New England										
Boston, Mass.	77.4	87.4	--	--	22.6	12.6	--	--	100.0	100.0
Middle Atlantic										
New York, N.Y.	75.8	71.6	--	--	21.2	22.7	3.0	5.7	100.0	100.0
South Atlantic										
Washington, D.C.	71.7	67.8	--	--	26.4	28.3	1.9	3.9	100.0	100.0
East North Central										
Chicago, Ill.	80.5	75.4	--	--	16.7	19.8	2.8	4.8	100.0	100.0
Detroit, Mich.	81.7	80.8	--	--	15.5	15.4	2.8	3.8	100.0	100.0
West North Central										
Minneapolis-St. Paul, Minn.	83.5	77.0	--	--	16.5	20.0	--	3.0	100.0	100.0
St. Louis, Mo.	81.4	76.0	--	--	16.5	20.1	2.1	3.9	100.0	100.0
Pacific										
Seattle, Wash.	78.3	74.9	--	--	17.9	20.4	3.8	4.7	100.0	100.0

Note: All data are as of the middle of July.

Source: Based on data from Platt's Oilgram Price Service, oil companies, Interstate Commerce Commission, Bulk Carrier Conference, Inc., American Petroleum Institute, and Commerce Clearing House.

Table 9-c-2

BREAKDOWN OF THE PRICE OF #6 FUEL OIL BY COST COMPONENTS
INDUSTRIAL

Census Region SMSA	Production 1960	Production 1971	Transportation 1960	Transportation 1971	Distribution 1960	Distribution 1971	Taxes 1960	Taxes 1971	Total 1960	Total 1971
					Dollars Per Barrel					
New England										
Boston, Mass.	2.26	4.02	--	--	0.66	0.58	--	--	2.92	4.60
Middle Atlantic										
New York, N.Y.	2.25	3.50	--	--	0.63	1.11	--	--	2.88	4.61
South Atlantic										
Washington, D.C.	2.25	3.50	--	--	0.83	1.46	--	--	3.08	4.96
East North Central										
Chicago, Ill.	2.85	3.88	--	--	0.59	1.02	0.10	0.25	3.54	5.15
Detroit, Mich.	3.21	3.62	--	--	0.61	0.69	--	--	3.82	4.31
West North Central										
Minneapolis-St. Paul, Minn.	2.98	3.62	--	--	0.59	0.94	--	--	3.57	4.56
St. Louis, Mo.	2.71	3.32	--	--	0.55	0.88	--	--	3.26	4.20
Pacific										
Seattle, Wash.	2.44	3.66	--	--	0.56	1.00	0.12	0.23	3.12	4.89

	Percent Distribution									
New England										
Boston, Mass.	77.4	87.4	--	--	22.6	12.6	--	--	100.0	100.0
Middle Atlantic										
New York, N.Y.	78.1	75.9	--	--	21.9	24.1	--	--	100.0	100.0
South Atlantic										
Washington, D.C.	73.1	70.6	--	--	26.9	29.4	--	--	100.0	100.0
East North Central										
Chicago, Ill.	80.5	75.3	--	--	16.7	19.8	2.8	4.9	100.0	100.0
Detroit, Mich.	84.0	84.0	--	--	16.0	16.0	--	--	100.0	100.0
West North Central										
Minneapolis-St. Paul, Minn.	83.5	79.4	--	--	16.5	20.6	--	--	100.0	100.0
St. Louis, Mo.	83.1	79.0	--	--	16.9	21.0	--	--	100.0	100.0
Pacific										
Seattle, Wash.	78.2	74.9	--	--	18.0	20.4	3.8	4.7	100.0	100.0

Note: All data are as of the middle of July.

Source: Based on data from Platt's Oilgram Price Service, oil companies, Interstate Commerce Commission, Bulk Carrier Conference, Inc., American Petroleum Institute, and Commerce Clearing House.

Table 9-d

BREAKDOWN OF RETAIL SERVICE STATION PRICE OF GASOLINE BY COST COMPONENTS

Cents Per Gallon

Census Region SMSA	Production 1960	Production 1971	Transportation 1960	Transportation 1971	Distribution 1960	Distribution 1971	Taxes[1/] 1960	Taxes[1/] 1971	Total 1960	Total 1971
New England										
Boston, Mass.	11.0	11.3	0.87	1.40	6.53	13.70	9.5	10.5	27.9	36.9
Middle Atlantic										
New York, N.Y.	11.0	11.3	0.87	0.73	9.33	14.97	10.7	12.9	31.9	39.9
Pittsburgh, Pa.	11.0	11.3	1.37	1.23	8.83	11.37	9.0	12.0	30.2	35.9
South Atlantic										
Atlanta, Ga.	11.0	11.3	0.83	0.51	7.87	13.69	11.2	12.4	30.9	37.9
Charlotte, N.C.	11.0	11.3	0.94	0.58	7.96	12.02	11.0	11.0	30.9	34.9
Washington, D.C.	11.0	11.3	1.02	0.67	7.88	13.93	10.0	11.0	29.9	36.9
East North Central										
Chicago, Ill.	11.0	11.3	0.79	0.79	11.31	13.31	9.8	13.0	32.9	38.4
Detroit, Mich.	11.0	11.3	1.44	1.44	10.66	7.16	10.8	12.0	33.9	31.9
Columbus, Ohio	11.0	11.3	1.31	1.31	8.59	14.29	11.0	11.0	31.9	37.9
East South Central										
Nashville, Tenn.	11.0	11.3	0.91	0.58	7.99	9.02	11.0	11.0	30.9	31.9
West North Central										
Minneapolis-St. Paul, Minn.	11.5	12.0	1.62	1.62	9.28	12.28	9.0	11.0	31.4	36.9
Omaha, Neb.	11.5	12.0	1.12	1.12	8.28	14.28	11.0	12.5	31.9	39.9
St. Louis, Mo.	11.5	12.0	0.48	0.43	9.42	15.47	8.5	9.0	29.9	36.9
West South Central										
Houston, Tex.	11.0	11.3	0.22	0.22	9.68	13.38	9.0	9.0	29.9	33.9
Tulsa, Okla.	11.5	12.0	0.24	0.24	10.66	14.16	10.5	10.5	32.9	36.9
Mountain										
Phoenix, Ariz.	10.5	11.5	1.10	1.10	6.30	14.30	9.0	11.0	26.9	37.9
Pacific										
Los Angeles, Calif.	10.5	11.5	0.12	0.12	9.28	14.28	10.0	11.0	29.9	36.9

Percent Distribution

Census Region SMSA	Production 1960	Production 1971	Transportation 1960	Transportation 1971	Distribution 1960	Distribution 1971	Taxes 1960	Taxes 1971	Total 1960	Total 1971
New England										
Boston, Mass.	39.4	30.6	3.1	3.8	23.4	37.2	34.1	28.4	100.0	100.0
Middle Atlantic										
New York, N.Y.	34.5	28.3	2.7	1.8	29.3	37.5	33.5	32.3	100.0	100.0
Pittsburgh, Pa.	36.4	31.5	4.5	3.4	29.2	31.7	29.8	33.4	100.0	100.0
South Atlantic										
Atlanta, Ga.	35.6	29.8	2.7	1.4	25.5	35.9	36.2	32.7	100.0	100.0
Charlotte, N.C.	35.6	32.4	3.0	1.7	25.8	34.4	35.	31.5	100.0	100.0
Washington, D.C.	36.8	30.6	3.4	1.8	26.4	37.8	33.4	29.8	100.0	100.0
East North Central										
Chicago, Ill.	33.4	29.4	2.4	2.1	34.4	34.7	29.8	33.9	100.0	100.0
Detroit, Mich.	32.5	35.4	4.2	4.5	31.4	22.5	31.9	37.6	100.0	100.0
Columbus, Ohio	34.5	29.8	4.1	3.5	26.9	37.7	34.5	29.0	100.0	100.0
East South Central										
Nashville, Tenn.	35.6	35.4	2.9	1.8	25.9	28.3	35.6	34.5	100.0	100.0
West North Central										
Minneapolis-St. Paul, Minn.	36.6	32.5	5.1	4.4	29.6	33.3	28.7	29.8	100.0	100.0
Omaha, Neb.	36.0	30.1	3.5	2.8	26.0	35.8	34.5	31.3	100.0	100.0
St. Louis, Mo.	38.5	32.5	1.6	1.2	31.5	41.9	28.4	24.4	100.0	100.0
West South Central										
Houston, Tex.	36.8	33.3	0.7	0.6	32.4	39.5	30.1	26.5	100.0	100.0
Tulsa, Okla.	35.0	32.5	0.7	0.6	32.4	38.4	31.9	28.5	100.0	100.0
Mountain										
Phoenix, Ariz.	39.0	30.3	4.1	2.9	23.4	37.7	33.5	29.0	100.0	100.0
Pacific										
Los Angeles, Calif.	35.1	31.2	0.4	0.3	31.0	38.7	33.4	29.8	100.0	100.0

1/ Major brand regular grade gasoline.

Note: Data are as of the first of July.

Source: Based on data from Platt's Oilgram Price Service, Oil and Gas Journal, Interstate Commerce Commission, oil companies, American Petroleum Institute, and Commerce Clearing House.

Table 9-e

BREAKDOWN OF THE PRICE OF COAL INTO COST COMPONENTS
INDUSTRIAL

Census Region SMSA	Dollars Per Ton										Percent Distribution									
	Production		Transportation		Distribution		Taxes		Total		Production		Transportation		Distribution		Taxes		Total	
	1960	1971	1960	1971	1960	1971	1960	1971	1960	1971	1960	1971	1960	1971	1960	1971	1960	1971	1960	1971
New England																				
Albany, N.Y.	4.62	7.35	5.12	6.62	--	--	--	--	9.74	13.97	47.4	52.6	52.6	47.4	--	--	--	--	100.0	100.0
Middle Atlantic																				
New York, N.Y.	7.10	13.78	5.12	6.62	--	--	--	--	12.22	20.40	58.1	67.5	41.9	32.5	--	--	--	--	100.0	100.0
Pittsburgh, Pa.	5.29	8.52	1.30	1.63	--	--	--	--	6.59	10.15	80.3	83.9	19.7	16.1	--	--	--	--	100.0	100.0
South Atlantic																				
Atlanta, Ga.	4.75	13.21	5.32	6.81	--	--	--	--	10.07	20.02	47.2	66.0	52.8	34.0	--	--	--	--	100.0	100.0
Charlotte, N.C.	4.84	13.40	4.34	5.86	--	--	0.09	0.19	9.27	19.45	52.2	68.9	46.8	30.1	--	--	1.0	1.0	100.0	100.0
East North Central																				
Chicago, Ill.	4.35	6.94	3.23	4.50	--	--	0.15	0.54	7.73	11.98	56.3	57.9	41.8	37.6	--	--	1.9	4.5	100.0	100.0
Detroit, Mich.	4.62	9.43	4.09	5.63	--	--	--	--	8.71	15.06	53.0	62.6	47.0	37.4	--	--	--	--	100.0	100.0
Columbus, Ohio	3.85	5.24	4.92	3.75	--	--	--	--	8.77	8.99	43.9	58.3	56.1	41.7	--	--	--	--	100.0	100.0
East South Central																				
Nashville, Tenn.	4.75	13.21	5.05	6.54	--	--	--	--	9.80	19.75	48.5	66.9	51.5	33.1	--	--	--	--	100.0	100.0
West North Central																				
Omaha, Neb.	4.31	4.87	1.53	4.92	--	--	--	--	5.84	9.79	73.8	49.7	26.2	50.3	--	--	--	--	100.0	100.0
St. Louis, Mo.	5.00	7.23	1.35	1.45	--	--	0.15	0.33	6.50	9.01	76.9	80.2	20.8	16.1	--	--	2.3	3.7	100.0	100.0
Mountain																				
Denver, Colo.	5.55	5.24	1.37	2.40	--	--	--	--	6.92	7.64	80.2	68.6	19.8	31.4	--	--	--	--	100.0	100.0
Salt Lake City, Utah	3.45	4.75	1.95	4.12	--	--	--	--	5.40	8.87	63.9	53.6	36.1	46.4	--	--	--	--	100.0	100.0

Note: Data are annual averages.

Source: Based on data from Coal Producers and Users, U.S. Bureau of Mines, and Commerce Clearing House.

Table 9-f-1

BREAKDOWN OF THE PRICE OF GAS INTO COST COMPONENTS RESIDENTIAL1/

Census Region SMSA	Cents Per Mcf										Percent Distribution									
	Production		Transportation		Distribution		Taxes		Total		Production		Transportation		Distribution		Taxes		Total	
	1960	1971	1960	1971	1960	1971	1960	1971	1960	1971	1960	1971	1960	1971	1960	1971	1960	1971	1960	1971
New England																				
Boston, Mass.	45.9	52.4	13.7	20.4	106.0	123.7	-0-	-0-	165.6	196.5	27.7	26.7	8.3	10.4	64.0	62.9	-0-	-0-	100.0	100.0
Middle Atlantic																				
New York, N.Y.	15.8	20.0	28.5	29.6	105.5	92.8	4.5	10.0	154.3	152.4	10.2	13.1	18.5	19.4	68.4	60.9	2.9	6.6	100.0	100.0
South Atlantic																				
Atlanta, Ga.	18.9	20.8	18.1	24.1	50.7	65.9	2.6	3.3	90.3	114.1	21.0	18.2	20.0	21.1	56.1	57.8	2.9	2.9	100.0	100.0
Charlotte, N.C.	15.7	20.3	10.2	20.4	91.0	68.9	3.8	3.3	129.7	112.9	12.1	18.0	14.8	18.1	70.2	61.0	2.9	2.9	100.0	100.0
Washington, D.C.	26.2	21.3	30.0	35.2	88.8	93.2	3.0	6.0	148.0	155.7	17.7	13.7	20.3	22.6	60.0	59.9	2.0	3.8	100.0	100.0
East North Central																				
Chicago, Ill.	15.4	18.6	16.5	17.6	75.5	77.4	6.0	8.9	113.4	122.5	13.6	15.2	14.5	14.3	66.6	63.2	5.3	7.3	100.0	100.0
Detroit, Mich.	16.8	20.2	22.9	22.1	46.3	45.7	2.5	2.1	86.5	90.1	19.4	22.4	24.2	24.5	53.5	50.8	2.9	2.3	100.0	100.0
East South Central																				
Nashville, Tenn.	36.4	38.7	6.6	7.1	70.8	71.8	3.4	3.5	117.1	121.1	31.0	31.9	5.6	5.9	60.5	59.3	2.9	2.9	100.0	100.0
West North Central																				
Minneapolis-St. Paul, Minn.	14.4	16.3	30.9	25.8	57.9	68.7	-0-	3.3	103.2	114.1	14.0	14.3	29.9	22.6	56.1	60.2	-0-	2.9	100.0	100.0
Omaha, Neb.	14.4	16.3	20.9	22.3	38.1	37.0	-0-	1.9	73.4	77.5	19.6	21.0	28.5	28.8	51.9	47.7	-0-	2.5	100.0	100.0
St. Louis, Mo.	16.1	15.3	15.2	29.9	65.4	64.7	1.9	5.9	98.6	115.7	16.4	13.1	15.4	25.9	66.3	55.9	1.9	5.1	100.0	100.0
Mountain																				
Phoenix, Ariz.	13.9	16.9	21.5	23.8	52.7	73.7	3.0	4.5	91.1	118.9	15.3	14.2	23.6	20.0	57.8	62.0	3.3	3.8	100.0	100.0
Pacific																				
Los Angeles, Calif.	13.9	16.9	17.7	20.5	70.4	61.4	3.1	9.0	105.1	107.8	13.2	15.7	16.9	19.0	67.0	57.0	2.9	8.3	100.0	100.0
San Francisco, Calif.	13.9	20.8	17.8	13.2	34.6	42.4	2.0	6.8	68.3	83.2	20.3	25.0	26.1	15.9	50.7	50.9	2.9	8.2	100.0	100.0

1/ Covers gas substitutable base load plus gas space heating.

Note: The production price is defined as that paid by the pipeline. In most instances it covers purchases in the field or from gasoline plants. In a few instances -- Boston, Chicago, Nashville and San Francisco -- it covers purchases from other pipelines in whole or in part. These prices are higher than the field prices since some transportation from the field is already included.

Source: Based on data from Form 2 Annual Reports by Interstate Pipelines to the FPC, gas utilities, Moody's Public Utility Manual, American Gas Association Rate Service, and Commerce Clearing House.

Table 9-f-2

BREAKDOWN OF THE PRICE OF GAS INTO COST COMPONENTS
SMALL COMMERCIAL

Census Region	Cents Per Mcf										Percent Distribution									
	Production		Transportation		Distribution		Taxes		Total		Production		Transportation		Distribution		Taxes		Total	
SMSA	1960	1971	1960	1971	1960	1971	1960	1971	1960	1971	1960	1971	1960	1971	1960	1971	1960	1971	1960	1971
New England																				
Boston, Mass.	45.9	52.4	13.7	20.4	159.5	185.1	-0-	-0-	219.1	257.9	20.9	20.3	6.3	7.9	72.8	71.8	-0-	-0-	100.0	100.0
Middle Atlantic																				
New York, N.Y.	15.8	20.0	28.5	29.6	119.2	118.1	5.0	12.0	168.5	179.7	9.4	11.1	16.9	16.5	70.7	65.7	3.0	6.7	100.0	100.0
South Atlantic																				
Atlanta, Ga.	18.9	20.9	18.1	24.1	53.1	61.4	2.7	3.2	92.8	109.5	20.4	19.0	19.5	22.0	57.2	56.1	2.9	2.9	100.0	100.0
Charlotte, N.C.	15.7	20.3	19.2	20.4	94.0	82.1	4.0	3.8	132.9	126.6	11.8	16.0	14.5	16.1	70.7	64.9	3.0	3.0	100.0	100.0
Washington, D.C.	26.2	21.3	30.0	35.2	77.0	85.6	2.7	5.7	135.9	147.8	19.3	14.4	22.1	23.8	56.6	57.9	2.0	3.9	100.0	100.0
East North Central																				
Chicago, Ill.	15.4	18.6	16.5	17.6	71.8	59.9	5.9	7.7	109.6	103.8	14.0	17.9	15.1	17.0	65.5	57.7	5.4	7.4	100.0	100.0
Detroit, Mich.	16.8	20.2	20.9	22.1	43.7	42.3	2.4	3.4	83.8	88.0	20.0	22.9	24.9	25.1	52.2	48.1	2.9	3.9	100.0	100.0
East South Central																				
Nashville, Tenn.	36.4	38.7	6.6	7.1	92.1	89.3	4.0	4.0	139.0	139.1	26.1	27.8	4.7	5.1	66.3	64.2	2.9	2.9	100.0	100.0
West North Central																				
Minneapolis-St. Paul, Minn.	14.4	16.3	30.9	25.8	47.4	57.9	-0-	3.1	92.7	103.1	15.5	15.8	33.4	25.0	51.1	56.2	-0-	3.0	100.0	100.0
Omaha, Neb.	14.4	16.3	20.9	22.3	45.6	31.4	-0-	1.8	80.9	71.8	17.8	22.7	25.8	31.7	56.4	43.7	-0-	2.5	100.0	100.0
St. Louis, Mo.	16.1	15.3	15.2	29.9	50.1	49.1	1.7	5.2	83.1	99.4	19.4	15.3	18.3	30.1	60.3	49.4	2.0	5.2	100.0	100.0
Mountain																				
Phoenix, Ariz.	13.9	16.9	21.5	23.8	31.6	39.0	5.9	3.1	72.9	82.8	19.1	20.4	29.5	28.8	43.3	47.1	8.1	3.7	100.0	100.0
Pacific																				
Los Angeles, Calif.	13.9	16.9	17.7	20.5	52.5	44.7	2.5	7.4	86.6	89.5	16.1	18.9	20.4	22.9	60.6	49.9	2.9	8.3	100.0	100.0
San Francisco, Calif.	13.9	20.8	17.8	13.2	31.1	36.9	1.6	6.4	64.4	77.3	21.6	26.9	27.6	17.1	48.3	47.7	2.5	8.3	100.0	100.0

Note: The production price is defined as that paid by the pipeline. In most instances it covers purchases in the field or from gasoline plants. In a few instances -- Boston, Chicago, Nashville and San Francisco -- it covers purchases from other pipelines in whole or in part. These prices are higher than the field prices since some transportation from the field is already included.

Source: Based on data from Form 2 Annual Reports by Interstate Pipelines to the FPC, gas utilities, Moody's Public Utility Manual, American Gas Association Rate Service, and Commerce Clearing House.

Table 9-f-3

BREAKDOWN OF THE PRICE OF GAS INTO COST COMPONENTS
LARGE COMMERCIAL

Census Region	Cents Per Mcf										Percent Distribution									
	Production		Transportation		Distribution		Taxes		Total		Production		Transportation		Distribution		Taxes		Total	
SMSA	1960	1971	1960	1971	1960	1971	1960	1971	1960	1971	1960	1971	1960	1971	1960	1971	1960	1971	1960	1971
New England																				
Boston, Mass.	45.9	52.4	13.7	20.4	78.5	108.9	-0-	-0-	138.1	181.7	33.2	28.9	9.9	11.2	56.9	59.9	-0-	-0-	100.0	100.0
Middle Atlantic																				
New York, N.Y.	15.8	20.0	28.5	29.6	89.0	81.3	4.0	9.2	137.3	140.1	11.5	14.3	20.8	21.1	64.8	58.0	2.9	6.6	100.0	100.0
South Atlantic																				
Atlanta, Ga.	18.9	20.8	18.1	24.1	53.0	59.1	2.7	3.1	92.7	107.1	20.4	19.4	19.5	22.5	57.2	55.2	2.9	2.9	100.0	100.0
Charlotte, N.C.	15.7	20.3	19.2	20.4	85.2	29.3	3.6	2.1	123.7	72.1	12.7	28.2	15.5	28.3	68.9	40.6	2.9	2.9	100.0	100.0
Washington, D.C.	26.2	21.3	30.0	35.2	49.1	53.4	2.1	4.4	107.4	114.3	24.4	18.6	27.9	30.8	45.7	46.7	2.0	3.9	100.0	100.0
East North Central																				
Chicago, Ill.	15.4	18.6	16.5	17.6	35.2	37.5	3.8	5.8	70.9	79.5	21.7	23.4	23.3	22.1	49.6	47.2	5.4	7.3	100.0	100.0
Detroit, Mich.	16.8	20.2	20.9	22.1	34.4	32.5	2.2	3.0	74.3	77.8	22.6	26.0	28.1	28.4	46.3	41.8	3.0	3.8	100.0	100.0
East South Central																				
Nashville, Tenn.	36.4	38.7	6.6	7.1	32.2	39.8	2.3	2.6	77.4	88.2	46.9	43.9	8.5	8.1	41.6	45.1	3.0	2.9	100.0	100.0
West North Central																				
Minneapolis-St. Paul, Minn.	14.4	16.3	30.9	25.8	24.7	46.4	-0-	2.7	70.0	91.2	20.6	17.9	44.1	28.3	35.3	50.9	-0-	2.9	100.0	100.0
Omaha, Neb.	14.4	16.3	20.9	22.3	26.0	13.7	-0-	1.3	61.3	53.6	23.5	30.4	34.1	41.6	42.4	25.6	-0-	2.4	100.0	100.0
St. Louis, Mo.	16.1	15.3	15.2	29.9	14.4	15.0	0.9	3.2	46.6	63.3	34.6	24.0	32.6	47.2	30.9	23.7	1.9	5.1	100.0	100.0
Mountain																				
Phoenix, Ariz.	13.9	16.9	21.5	23.8	9.8	15.0	3.2	2.2	48.4	57.9	28.7	29.2	44.4	41.1	20.3	25.9	6.6	3.8	100.0	100.0
Pacific																				
Los Angeles, Calif.	13.9	16.9	17.7	20.5	32.7	30.8	1.9	6.1	66.2	74.3	21.0	22.7	26.7	27.6	49.4	41.5	2.9	8.2	100.0	100.0
San Francisco, Calif.	13.9	20.8	17.8	13.2	24.7	26.3	1.7	5.4	58.1	65.7	23.9	31.7	30.7	20.1	42.5	40.0	2.9	8.2	100.0	100.0

Note: The production price is defined as that paid by the pipeline. In most instances it covers purchases in the field or from gasoline plants. In a few instances -- Boston, Chicago, Nashville and San Francisco -- it covers purchases from other pipelines in whole or in part. These prices are higher than the field prices since some transportation from the field is already included.

Source: Based on data from Form 2 Annual Reports by Interstate Pipelines to the FPC, gas utilities, Moody's Public Utility Manual, American Gas Association Rate Service, and Commerce Clearing House.

Table 9-f-4

BREAKDOWN OF THE PRICE OF GAS INTO COST COMPONENTS
INDUSTRIAL[1/]

Census Region SMSA	Cents Per Mcf										Percent Distribution									
	Production		Transportation		Distribution		Taxes		Total		Production		Transportation		Distribution		Taxes		Total	
	1960	1971	1960	1971	1960	1971	1960	1971	1960	1971	1960	1971	1960	1971	1960	1971	1960	1971	1960	1971
New England																				
Boston, Mass.	45.9	52.4	13.7	20.4	50.1	106.6	-0-	-0-	109.7	179.4	41.8	29.2	12.5	11.4	45.7	59.4	-0-	-0-	100.0	100.0
Middle Atlantic																				
New York, N.Y.	15.8	20.0	28.5	29.6	83.1	80.6	-0-	-0-	127.3	130.2	12.4	15.4	22.4	22.7	65.2	61.9	-0-	-0-	100.0	100.0
South Atlantic																				
Atlanta, Ga.	18.9	20.8	18.1	24.1	25.9	44.4	-0-	-0-	62.9	89.3	30.0	23.3	28.8	27.0	41.2	49.7	-0-	-0-	100.0	100.0
Charlotte, N.C.	15.7	20.3	19.2	20.4	85.0	29.2	1.2	0.7	121.1	70.6	13.0	28.7	15.8	28.9	70.2	41.4	1.0	1.0	100.0	100.0
Washington, D.C.	26.2	21.3	30.0	35.2	40.6	44.9	-0-	-0-	96.8	101.4	27.1	21.0	31.0	34.7	41.9	44.3	-0-	-0-	100.0	100.0
East North Central																				
Chicago, Ill.	15.4	18.6	16.5	17.6	28.6	31.8	3.4	5.4	63.9	73.4	24.1	25.3	25.8	24.0	44.8	43.3	5.3	7.4	100.0	100.0
Detroit, Mich.	16.8	20.2	20.9	22.1	19.3	22.9	-0-	-0-	57.0	65.2	29.5	31.0	36.7	33.9	33.8	35.1	-0-	-0-	100.0	100.0
East South Central																				
Nashville, Tenn.	36.4	38.7	6.6	7.1	14.3	20.6	-0-	-0-	57.2	66.4	63.5	58.3	11.5	10.7	25.0	31.0	-0-	-0-	100.0	100.0
West North Central																				
Minneapolis-St. Paul, Minn.	14.4	16.3	30.9	25.8	22.9	34.6	-0-	-0-	68.2	76.7	21.1	21.3	45.3	33.6	33.6	45.1	-0-	-0-	100.0	100.0
Omaha, Neb.	14.4	16.3	20.9	22.3	n.a.	9.4	n.a.	-0-	n.a.	48.0	n.a.	34.0	n.a.	46.4	n.a.	19.6	n.a.	-0-	100.0	100.0
St. Louis, Mo.	16.1	15.3	15.2	29.9	5.3	5.1	-0-	2.5	36.6	52.7	44.0	28.9	41.5	56.7	14.5	9.7	-0-	4.7	100.0	100.0
Mountain																				
Phoenix, Ariz.	13.9	16.9	21.5	23.8	9.6	10.7	-0-	2.0	45.0	53.4	30.9	31.7	47.8	44.6	21.3	20.0	-0-	3.7	100.0	100.0
Pacific																				
Los Angeles, Calif.	13.9	16.9	17.7	20.5	31.0	29.2	-0-	-0-	62.6	66.6	22.2	25.4	28.3	30.8	49.5	43.8	-0-	-0-	100.0	100.0
San Francisco, Calif.	13.9	20.8	17.8	13.2	23.8	25.3	-0-	-0-	55.5	59.3	25.0	35.1	32.1	22.2	42.9	42.7	-0-	-0-	100.0	100.0

1/ Firm.

Note: The production price is defined as that paid by the pipeline. In most instances it covers purchases in the field or from gasoline plants. In a few instances -- Boston, Chicago, Nashville and San Francisco -- it covers purchases from other pipelines in whole or in part. These prices are higher than the field prices since some transportation from the field is already included.

Source: Based on data from Form 2 Annual Reports by Interstate Pipelines to the FPC, gas utilities, Moody's Public Utility Manual, American Gas Association Rate Service, and Commerce Clearing House.

Reviewer Comments

n/e/r/a

March 28, 1974

The Energy Policy Project
1776 Massachusetts Avenue, N.W.
Washington, D.C. 20036

Attention: Mr. Sheridan

Gentlemen:

You have suggested that, as one of the reviewers of your report, I comment on the final draft of the Foster Associates study entitled "Retail and Wholesale Prices for Primary and Secondary Energy Sources in the United States—1960-1970," with a 1973 addendum. Obviously time did not permit me, nor was it feasible, to concern myself with the data collection process. This appears to have been handled in a thoroughly competent fashion, and the results as reported will no doubt be of considerable value in analyzing this period. The comments I have (some of which I made before) are quite limited. For what they are worth, here they are.

PART I

1. The statement on *page 2* that "the rise in energy prices was *not* a unique price development" is not in accordance with *my* reading of the data.

2. The statement at the end of the overview (page two) does not say why. The reason is obvious—as the base cost of energy rose, by a common unit amount (viz. cents per Mcf) to each user, it constituted a greater proportionate increase in the low price sales than in the high price sales.

3. *Pages 18, et seq.* This section suffers from a lack of discussion of the FPC's recent pricing policies, whereby *new* sales are currently being made at two to three times the last figure quoted on Table 3-4 a very serious omission if we are to understand current price relationships.

4. *Pages 33, et seq.* This section does not include a discussion of the place of the "posted price" in the "costing" of foreign oil shipped to the United States. Thus, raising this price is, in important part, a *tax* measure on the part of the producing countries, taxes which may be offset against U.S. income taxes. The net effect, to the consumer and to the U.S. Treasury, cannot be derived from the commentary, hence the reader is likely to obtain a distorted impression of the impact of the posted price dislocations of recent years.

I have more trouble with Part II.

1. *Pp. 60, et seq.* The conversion factors used to find electric use equivalents to assumed gas usage were derived from gas industry sources, apparently unchecked, with the EEI or any such source, and put gas in its most favorable light, competitively speaking—a too favorable light in my judgment. I fear that the comparisons shown in Tables 3-1, 3-2, and 3-3 may be misleading as a result, as well as the commentary thereon.

2. *Same Section and Section D, page 110.* There is reference continually herein to "promotional" rates, now somewhat of a term of obloquy. Actually electric and gas rates may be "promotional" because their declining slopes reflect the lower unit costs of *distributing* larger amounts of energy per customer. Thus, they properly follow *cost,* and are promotional only as a consequence. There have, of course, been some genuinely promotional rates offered but they are in the distinct minority. Again I think the commentary suffers by this confusion.

3. *Table 3-14 and Commentary, page 85.* In the period 1969 to January 1973, No. 6 fuel oil prices did rise markedly. This was, however, *primarily* a phenomenon resulting from a lag in constructing desulfurization facilities, with a resulting shortage of low-sulfur product; a circumstance which can be surmised from the table, but is not, in my judgment, adequately commented on.

4. *Pages 93, et seq. Section A.* This section attempts to breakdown total gas, oil and electric costs into reasonably homogeneous components (production, transportation, etc.) and then makes certain intrafuel comparisons as to the variations. Of course, the proportions will vary, because of the

widely varying physical characteristics of these sources of energy. It doesn't contribute to clarity of analysis to make these comparisons as though there were some significance therein, when they only reflect these physical characteristics.

These thoughts are submitted for what significance you think they have.

Sincerely,

Charles H. Frazier
Director, Philadelphia Office
National Economic Research Associates, Inc.
Consulting Economists